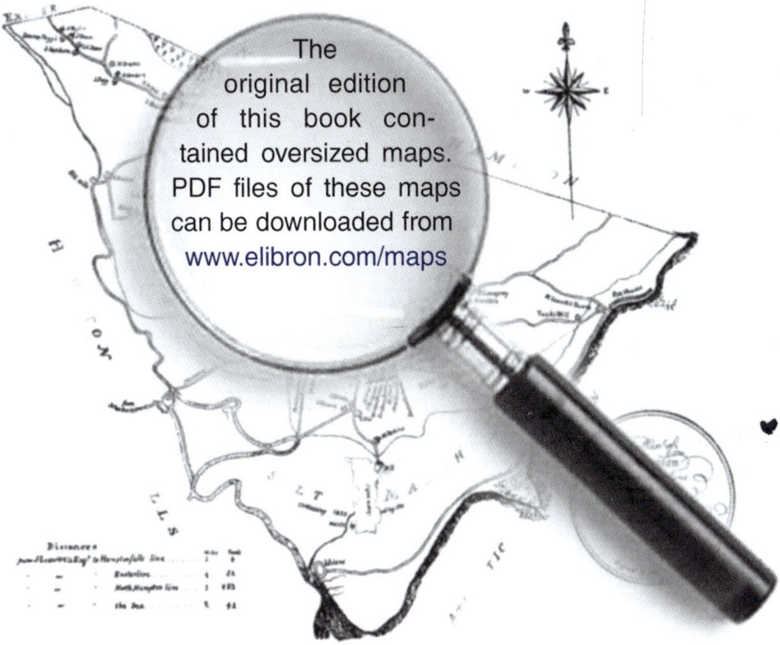
The
original edition
of this book con-
tained oversized maps.
PDF files of these maps
can be downloaded from
www.elibron.com/maps

HERBERT WOOD

THE

# SHORES OF LAKE ARAL

Elibron Classics
www.elibron.com

Elibron Classics series.



ISBN 1-4212-6887-6 (paperback)
ISBN 1-4212-6886-8 (hardcover)

This Elibron Classics Replica Edition is an unabridged facsimile of the edition published in 1876 by Smith, Elder and Co., London.

# LAKE ARAL.

THE

# SHORES OF LAKE ARAL.

BY

HERBERT WOOD,

MAJOR, ROYAL ENGINEERS:
FELLOW OF THE ROYAL AND OF THE IMPERIAL RUSSIAN GEOGRAPHICAL SOCIETIES
CORRESPONDING MEMBER OF THE SOCIETY OF GEOGRAPHY OF GENEVA.

'*We were the first that ever burst*
*Into that silent sea.*'—COLERIDGE, *Ancient Mariner.*

With Maps.

LONDON:
SMITH, ELDER, & CO., 15 WATERLOO PLACE.
1876.

TO

CHARLOTTE AND ATHENAÏS DE BODISCO

This Book is Inscribed

BY

THEIR OBLIGED FRIEND AND SERVANT,

THE AUTHOR.

# PREFACE.

*'Le continent de l'Asie, à ses extrémités du sud et du nord, offre deux vastes empires soumis à la domination européenne. C'est comme deux mondes. . . . C'est le monde anglo-indou et le monde russe-sibérien, aussi différents par la nature des climats et des productions que par l'état de culture, la richesse et la densité de leur population. Ces deux grandes masses ou divisions ne communiquent depuis des siècles, que par les basses régions de la Bactriane, je pourrais dire par la dépression du sol qui entoure l'Aral et le bord oriental de la Caspienne entre Balkh et Astrabad, comme entre Tashkent et l'isthme de Troukhmènes. C'est une bande de terrains en partie très-fertiles, à travers laquelle l'Oxus a tracé son cours. . . . C'est le chemin de Delhy, de Lahore et de Kaboul à Khiva et à Orenbourg, la grande route par laquelle la puissance mongole a pénétré jadis dans l'Inde . . . Les régions que nous signalons ici peuvent être envisagées comme voie de communications paisibles ou d'invasions hostiles et progressives par stations.'*—HUMBOLDT, 'Asie Centrale.'

THE country spoken of in this passage, and included in the geographical expression West Turkestan, presents perhaps the greatest number of interesting points that could be found collected in any one single locality for the consideration of mankind. From the remotest ages, across these regions led the track which was followed by irruptions of Turanians, as they continually advanced from the far north-east, and pushed primitive Aryan races before them into Europe. From the south-west Egyptians, Assyrians or Medes, Persians, Greeks or Arabs, have passed as conquerors at different epochs into the regions watered by the Oxus and Jaxartes; while the creeds of Zoroaster and Buddha and Mohammed have successively agitated the breasts of their populations. Though the earliest historical records inform us how solitude and sterility were the marked characteristics of North-western Asia, there are certain indications of a previous local civilisation and culture, which were overwhelmed by the devastating torrent of perpetually encroaching barbarism; and such dim testimonies to a pre-historic state gain form and substance daily from observation of the signs that are graven on the face of Nature herself. These promise revelations fully as important perhaps as those which have been deciphered on the monuments of Egypt and of Assyria; for it would be difficult to define the limits of the lesson that is involved in a thorough study of the desolation of the Khwarezmian deserts, and the past action of mankind by which it was caused.

Besides the general human interests that are presented by the present aspects of West Turkestan, and connected more specially with bygone times, there are others which bear on the future; which are comprised in the extension of Russian rule in one part of Asia and the continued development of British power in another part. The eventual dethronement of degrading superstition, and the increase of material well-being among the millions of India, find their counterpart in the preliminary work of civilisation that lies before Russia in the Oxus countries—a work which involves the re-establishment of disturbed terraqueous harmonies, and the restoration of immense wildernesses to productiveness and fitness for ministering to the wants of mankind.

Having been graciously accorded permission by H. I. H. the Grand Duke Constantine, the President, to accompany an expedition sent under the auspices of the Imperial Russian Geographical Society to examine the Amúdarya, I had the opportunity of passing several months of the year 1874 in the countries round Lake Aral. In the following pages I have submitted to my readers a *résumé* of the information relating to these localities, whose future development has been regarded more exclusively from the stand-point of the change that may be effected in their superficial geography, and that is urgently called for by the social and political necessities of Russia, in presence of the impoverished resources of West Turkestan. I have endeavoured to present a faithful picture of regions whose sterile character thrusts such impediments in the way of Central Asian progress; while the hydraulic phenomena presented by the rivers have been treated of in some detail, since their study throws light on the curious physical history of the Aralo-Caspian region. Though I can but express a regret that these difficult and interesting problems have not here found a more competent expositor, a solution has been submitted which is founded on personal observation, as well as on the more complete knowledge recently acquired by Russian exploration. I should add that great obligations are due to Professor M. J. De Goeje, of the University of Leyden, who in a recent work on the ancient channel of the Oxus,[1] very considerately published several important passages from early Arabian geographers, in order that the information obtained by the Amú

[1] *Das alte Bett des Oxus.* Leiden: E. J. Brill. 1875.

expedition of 1874, might be more easily and expeditiously utilised. It is not to be supposed that all the difficulties attaching to such much debated questions are thoroughly disposed of in the following pages.[1] All I can hope is that some contribution will be found to have been made to previously existent knowledge, while firmer ground has been indicated from whence a fresh advance may be made towards the ultimate solution of the problems involved.

The Russian annexation of the Amúdarya territory promises to mark the commencement of an entirely new era in the history of Turkestan; for though the control of the river may not carry with it all those advantages of a great line of water-communication which Burnes some forty years ago predicted, and which Russia herself anticipated, it is not the less certain that the moral and material regeneration of Central Asia are dependent on the stream of the ancient Oxus. In semi-tropical regions a great river may be employed for more important ends than those of navigation, and flow-

[1] This remark applies more particularly to the supposed former passage of the Oxus waters by the bed of the Attrek, which is spoken of in Chapter XXII.

ing as they now do, northwards to Lake Aral, along the higher limits of the Khwarezmian plain the waters of the Amú may be artificially conducted down the slope of the country by canals of irrigation, towards the drainage line which was followed by the river in antique times, to its outlet in the Caspian Sea. In this way a belt of desert country may be reclaimed, and settled, industrious populations may once more be established upon broad lands, from whence the ancestors of European peoples were expelled by barbarian hordes. Impassable solitudes may become a great commercial highway into the now isolated regions of Central Asia, and the cause of progress be thus importantly served. The first steps have already been taken for the diffusion of light in these dark places of the earth, for Humanity owes to Russia the cessation of brigandage and slavery, which from the earliest historical times have been the scourges of the Oxus countries. But a more tangible reward would result from the restoration of the wilds of Turkestan. It would be for the benefit of Russia more particularly, were the deserts south of Khiva changed into productive cotton-fields, and into a habitable territory across which the ancient silk trade, that was destroyed by Mohammedan fanaticism, might once more be established between Europe and Central Asia.

Such, speaking generally, appears to be the promise of the study contained in the following pages; and though opinion in England leans daily more and more to the better part of sympathising with Russia in her struggle with Asiatic barbarism, to the British people above all others must her progress in the East be one of the most momentous of questions. How the contact of the two great European Powers that must inevitably take place in Asia, may be affected by a possible reclamation of the deserts round Khiva, is consequently an enquiry which immediately presents itself for discussion. I should be as presumptuous as I am little qualified, were I to attempt to treat this subject; but there are some weighty considerations that show such a project would be the opposite of detrimental to the interests of British India.

In the first place, it is to be noted that the possible development would take place in localities at a considerable distance from those deserts round Merv, where recent Russian movements have threatened to precipitate a discussion of the difficult boundary question between the two nations. It might not perhaps be entirely unreasonable to think, that the possession of a belt of cultivated and inhabited territory, extending directly from the Caspian to the valleys of the Zarafshán and Syrdarya, would nullify the need now felt by Russia, of re-establishing the old Seric commercial route from the lower Attrek along the northern frontiers of Khorassan. Thus the question of Russian boundaries on the Perso-Affghan frontier might be ultimately solved in a satisfactory way to both Powers.

The opportunity moreover which would be afforded by the reclamation of the Khwarezmian deserts for the settlement of the most barbarous of the Turkoman tribes, would remove one great cause of uneasiness which is common to the statesmen both of England and of Russia ; while the civilising current passing through the reclaimed country would infallibly act upon neighbouring countries, whose geographical position renders them inaccessible to peaceful influences from Hindustan. Nor is it improbable that the national unity and sympathy with progress which British statesmen are endeavouring to realise among the chiefs and tribes on the north-west frontier of India, would receive a felicitous impulse from the enormous material benefits which promise to accrue to

Central Asian peoples. The hitherto disquieting vagueness of Russian intentions in South-Western Asia would vanish before a pronounced policy of pacification and the application of her energies to the commercial development of the Central Asian Khanates; while the stereotyped tradition which is glanced at in the passage from Humboldt heading these prefatory remarks, and which has regarded the regions of the Oxus as mere stepping-stones for more southerly invasions, would, if not entirely inverted, be very materially modified, along with the transformation of physical aspects that have hitherto rendered these deserts unproductive and uninhabitable.

The fatuous dream that history might repeat itself, and that the British India of to-day might be overrun as the Gangetic plains formerly were by Mongol hordes, has all but vanished before the more sober and practical ideas of the present time. Russians themselves may be pardoned the contemplation of such a project, since Englishmen allowed this nightmare to possess them; though the calmer heads of our own people and the wiser spirits who have led Russia in the march of progress may well have regretted the perversity of the mass of mankind. At the present important epoch in the history of England in Asia it is a fortunate circumstance perhaps, that several intelligent Russians should be travelling in Hindustan; for whatever shortcomings their critical eyes may note, they will at least be able to enlighten their countrymen on the true bases of Anglo-Indian supremacy. Weighed down by gigantic armaments, the result of apprehension on the European frontier, the Russian population commences to realise that a military restlessness which daily and appreciably adds to financial burdens must give place to a policy of commercial development in Central Asia: while in danger of being entirely supplanted in the corn trade by America, the encouragement of cotton and silk production in regions specially adapted for such industries is more urgently pressed upon their notice, as a sheet-anchor in possible future commercial storms. The magnitude and practical difficulties attaching to such a project will doubtless demand the utmost and undivided energy of Russia for ensuring its success. But such a recurrence to the original commercial schemes of Peter the Great must inevitably accentuate a peace policy in the East, as well as necessitate a close approximation to England's earnest desire for the undisturbed rest which will assure the continuous development of European civilisation in Asia.

In conclusion, therefore, there are good grounds for thinking that a reclamation of the portion of the deserts which lie between the Caspian Sea and the Amúdarya would be to the benefit of both the Powers whose influence is paramount in Asia. In the succeeding pages it will be shown that such a restoration can be made; for the facts which are adduced, together with the information which has been collected from historical records, can scarcely fail to lead the reader to this conclusion. And interesting as such a result may be, its importance perhaps is still greater, for it might serve as a basis of arrangements for ensuring that the future contact of Russian and Anglo-Indian frontiers shall take place in the peaceful and ordinary manner of civilised Christian nations.

H. W.

# CONTENTS.

## CHAPTER I.

### SAMARA TO ORSK.



## CHAPTER II.

### SAMARA TO ORSK—*continued.*



## CHAPTER III.

### ORSK TO IRGEEZ.

## CHAPTER VII.

THE RUSSIANS ON THE SYR-DARYA—*continued.*



## CHAPTER VIII.

THE SYR-DARYA.



## CHAPTER IX.

THE BASIN OF LAKE ARAL.

## CHAPTER X.

### THE ARALO-CASPIAN SEA.



## CHAPTER XI.

### THE CASPIAN OF HISTORY.



## CHAPTER XII.

### THE CENTRAL OUTLET OF THE AMÚ-DARYA.

## CHAPTER XIII.

### THE LOWER AMÚ-DARYA.



## CHAPTER XIV.

### THE LOWER AMÚ-DARYA—*continued.*



## CHAPTER XV.

### THE CENTRAL AMÚ SWAMPS.

## CHAPTER XVI.

### CHARACTERISTICS OF THE AMÚ.



## CHAPTER XVII.

### THE AMÚ IN KHIVA.



## CHAPTER XVIII.

### THE LAST CHANGE IN THE OXUS.

## CHAPTER XIX.

### RUSSIA ON THE AMÚ.



## CHAPTER XX.

### THE BANKS OF THE AMÚ.



## CHAPTER XXI.

### THE BANKS OF THE AMÚ—*continued.*



## CHAPTER XXII.

### THE ANCIENT OXUS.

## CHAPTER XXIII.

### KHWAREZM AND THE JYHÚN.



## CHAPTER XXIV.

### KHWAREZM AND THE JYHÚN—*continued.*



## CHAPTER XXV.

### THE STEPPES OF TURKESTAN.

## CHAPTER XXVI.

### A RIDE ACROSS THE KIZZELKOOM.



## CHAPTER XXVII.

### A RIDE ACROSS THE KIZZELKOOM—*continued.*



---

## MAPS.

## *Errata.*

| | | | |
|---|---|---|---|
| Page | 36, | line | 20, *for* of Syr *read* of the Syr |
| ,, | 68, | ,, | 3, *delete* north of the Syr |
| ,, | 224, | ,, | 20, *for* Reynolds *read* Thompson |
| ,, | 334, | ,, | 2, *delete* east |
| ,, | 343, | ,, | 14, *for* SSW. *read* SSE. |

THE

# SHORES OF LAKE ARAL.

## CHAPTER I.

### SAMARA TO ORSK.

The river Volga and its importance to Russia—Great towns upon its banks—Their growing wealth—A riverside morning scene—The route into Asia—Orenburg—Ural Cossacks—Progress in the east of Russia—Ural valley to Orsk—Characteristics of population, towns, and villages from the Volga eastwards to Orsk—Spartan simplicity in Russia.

THE European railways, which run half-way across the four thousand miles of plain extending from Holland to Chinese Zungaria, come to an end for the present on the Volga. From Samara, on the Volga's left bank, commences the postal service, which carries the traveller into Turkestan. It was in May, when the snows of winter had all but disappeared, and when the waters of the great river and its affluents rolled along in high flood, that I was first at Samara. The town is also situated on the river of the same name where it joins the Volga after flowing down from the steppes on the east, and the Samara was submerging the low-lying buildings at the time of my visit. The drainage, which at the close of winter is poured into the Volga by its large and numerous tributaries, is something enormous, as may be gathered from the fact that in spring its surface is about thirty feet higher than the low-water level, while at some distance below Samara the ordinary breadth of two miles is increased to fifteen! Its discharge at this epoch is indeed stated to be about one seventh of the whole quantity of water which flows in all the rivers of Europe put together.

It would be interesting to trace what the civilisation of Europe in general, and that of Russia in particular, owes to the existence of this great stream, which at the present day is navigated by some six hundred steam-boats, and perhaps four times that number of cargo-flats. The part it has filled in the development of the empire has been a most important one, and as is natural, Russians are as proud of the Volga as Americans can be of the Mississippi. Martin Chuzzlewit on his way to Eden was indeed strongly suggested to me, when a Russian, influenced by a somewhat uneasy patriotism, affirmed that the Indian Ganges must be *très peu de chose* in comparison with the old Tartar stream; whose excess waters, however, might perhaps be advantageously utilised, like those of the Ganges, for the prevention of droughts, instead of being allowed to run uselessly away and be lost in evaporation from the Caspian Sea.

Samara, Saratoff, and other young giant cities on the middle course of the Volga, owe their birth and their rapid growth to the fertility of the provinces in which they are situated. In addition to the advantages they derive from their riverside position, the prosperity of these places, wherein the agricultural produce of the surrounding country is collected, has been aided by that development of communications which has made Moscow the centre of a spider's web of railways. Notwithstanding their increasing size and wealth, they do not present a very inviting aspect on a close inspection; for lacking that halo of historic memories which adds so great a charm to Moscow, the tawdry stucco architecture of their more pretentious buildings, which are scattered among the mean wooden houses of the ill-paved streets, impresses a stranger disagreeably. In one respect, however, they quite satisfy æsthetic requirements, for I know of few more glorious spectacles than that presented by these great Volga towns, when the rays of the rising sun pierce through their tender, blue, overhanging haze, and fleck their metal domes and spires with dazzling light. The glittering cupolas and pinnacles, rising high above a sea of particoloured roofs, whose gay tints tone into the neutral breadth of the surrounding steppe, form a scene at once delightful to the eye and pleasing to the inner sense of the beholder. Not seldom have I thought that such a picture was no inapt image of Russia herself, and her varied social and intellectual contrasts—of her bright and eminent realities of the present, her immense and mysterious possibilities of the future.

The curiosity of the passing traveller is soon satisfied by the towns of the Volga, and nothing will delay his speedy start from Samara for Asia, but the necessity of purchasing and packing his *tarantasse*, or posting carriage, in which the journey must be made. The Government keep up an establishment of horses for hire along the route, and previous to the end of 1874 every traveller into Turkestan had to procure a special order for these; at the present time such an order is not necessary as far as Orsk, to which place, therefore, all the world may travel, if so inclined.

Of the one thousand miles separating the Volga from the Syrdarya, the first quarter of the distance lies across the undulating and fertile plains which are traversed by the river Samara; during the ascent of whose valley the landscape assumes a continually bleaker appearance, amounting almost to wildness at the crest of the watershed, where the high ground running down from the north is reached. Fifty miles of descent from this somewhat stony locality takes the traveller to the river Ural, whose wooded banks contrast agreeably with the bare open plains which have been traversed so far. The Ural valley is ascended for a few miles, and the route then reaches the Sakmara, which flows from the east in a course north of and nearly parallel to that of the larger river of the Ural. The Sakmara is crossed just above the confluence of the two rivers; and in the fork between them, which is occupied by a smooth grassy plain, is situated, at a distance of about five miles, the city of Orenburg.

Orenburg, 'the city in the east,' is the frontier station from which for more than a century the nomadic populations were controlled, and the Russian advance into Central Asia was directed. Besides some large Government buildings, it possesses a fine caravanserai for the accommodation of Asiatic merchants and travellers, as well as a mosque, with its accompanying minaret, for the use of the neighbouring Kirghiz and Bashkir populations; and at the east end of the city, upon the high bank of the river Ural, are some public gardens, overlooked by the residences of the higher officials.

The Ural river, which runs westwards from Orsk by Orenburg, turns at Uralsk southwards to an outlet in the Caspian. It was formerly called the Yaïk, but this name was changed at the end ot the last century, in consequence of a rebellion of the Cossacks cantoned along its banks. These bigoted adherents of the old Orthodox Greek faith, having become dissatisfied at various social reforms, which had been instituted by Peter the Great and continued by his successors, and having been promised the repeal of obnoxious laws (such as the one which ordered the shaving of the beard), elected to cast in their lot with the pretender Pug acheff, in his attempt in 1773–74, to possess himself of the throne of the Czars. Still stands on the left bank of the Sakmara, the tower from whose summit Pugacheff watched the progress of the siege operations, which he undertook against Orenburg. After the failure of this enterprise, and after the capture and execution of the rebel chief, the Yaïk was ordered by Imperial ukase to be henceforth called the Ural; very much to the disgust of the Cossacks, whose descendants, to this day, employ the old name among themselves. The tenacity with which they cling to old ideas and traditions is almost worthy of Hindoos, and has been lately exemplified on the promulgation of the new law of obligatory service in the Russian army; for before this could be enforced among the Ural Cossacks, the Government were compelled to use armed pressure, and to deport several of the more influential among them to Siberia.[1] Holding their land by military tenure, they could not, of course, object to the principle of the new law, but they desired, as they had done in the past, to settle the details of their obligations among themselves; and this the Government did not concede, since the old system of relieving the rich from the duty of military service, would have been perpetuated by the admission of poorer and purchased substitutes.

Though Orenburg, in a political point of view, has lost much of its importance, since the establishment of the Government of Turkestan, it remains

[1] It appears their destination has been changed to Turkestan.

the head-quarters for the administration of Kirghiz and Bashkir affairs. Its commercial importance, which has always been considerable, from the circumstance of its being a depôt for Asiatic trade, and from its propinquity to the metalliferous districts of the Ural, will receive a further development on the opening of the line of railway which is now being constructed between it and Samara. A second line for mineral transport has recently been projected from Orenburg to the coal deposits in the lower Ural valley, and this energy in the extension of communications will give an immense impetus to commercial industry in districts which a few years ago were the *Ultima Thule* of Europe in the East.

Continuing eastwards from Orenburg, the route into Asia passes for two hundred miles to Orsk, up the valley of the Ural river, along its northern bank; on which were established, more than a century ago, the frontier Cossack outposts, for overawing the Kirghiz of the steppes, and for protecting and extending the commerce of Russia with Central Asia. The sparsely-wooded lands in this valley are generally cultivated. The uncultivated spots are the southern slopes of the Irendik hills, which run down from the north to the river, and are surmounted by an exposed grassy plateau which is crossed before descending into the valley again, some twenty miles from Orsk. Among these highlands are picturesque glens, where brawling mountain streams descend between wooded banks, and where the hut villages are redeemed by the scenery from that ugliness which the monotony of Russian plains makes so noticeable.

Between the parallel Irendik and Ural chains, the river Ural flows directly southwards to Orsk, where it turns sharply to Orenburg, instead of continuing to an outlet in Lake Aral, from which basin it is cut off by a range of hills extending along its left bank. Orsk is a little town, situated on the Asiatic side of the river, at whose right bank trees cease entirely. When the traveller reaches the ferry, many women and children, with boughs and green fodder for their goats, will cross with him. This little place is merely a much smaller edition of Orenburg, though its population has probably a still larger infusion of Kirghiz and of other Eastern blood. The valley traversed by the middle course of the Ural river has, during the past two centuries, been successively occupied by Nogai Tartars, Bashkirs, and Calmuks, and though the population is Russianized, these peoples have left very noticeable traces in the quaint pink and yellow-faced children of the villages, which are passed through in travelling from Orenburg.

The country from the Volga to Orsk may be described generally as fit for agriculture, more especially towards its western extremity, but it is also susceptible of improvement as population gets denser. The steppes through which the Samara river runs are magnificent cornlands, but here in 1873 prevailed scarcity which amounted to famine, though the actual distress was probably exaggerated, if, as was stated, prices at Orenburg were generally higher than in the Samara district itself. Along the same distance the villages are large and numerous, though only one place, called Bouzoulouk, occurs, near the head of the Samara valley, which has any pretensions (Orenburg of course excepted), to the name of a town. Towns and villages in this part of the world have much the same look, and differ only in size, for both are built in the same monotonous rows of shabby wooden houses, bordering broad unpaved avenues, which, for want of another word, must be called streets; though towns possess two or more churches, and a few brick buildings, devoted to the uses of Government, and have also some shops, whose exteriors are covered with gigantic daubs, picturing the various articles which may be bought within. A chief feature in every collection of buildings aspiring to the dignity of a town, is the lighthouse-looking watchtower, from the little gallery surrounding whose summit a sharp lookout is kept for fires. To guard against these there is sometimes a rule against smoking in the streets, and every Russian carefully treads upon the burning end of a cigarette, instead of throwing it carelessly away, as more civilised people do in the west of Europe. Towns and villages are entered at once from the bare surrounding steppes, without passing through any suburbs, though in the environs of Samara are some pretty *châlets*, which have small gardens around them. *Traktirs*, or dram shops, are found in more than suffic

towns and villages; and the former have generally at least one *gostonnitza* or hotel, of whose comforts and decencies the less said the better. A Russian travelling into Asia, leaves the great number of the niceties of civilised life behind, with his heavy baggage, at St. Petersburg, and carries but very few with him across the Volga. Becoming thus more and more of an Asiatic as he goes eastward, he does not feel the want of, and consequently does not demand, those surroundings which many other Europeans regard as mere necessaries, and whose absence east of the Volga are consequently very remarkable. The modest kit of an officer while yet a thousand miles away from the scene of his future service in Turkestan, is big with the promise of that Spartan simplicity which is suitable to a warlike nationality, and which the late Sir Charles Napier so forcibly urged on the Sybarite legions under his command in India. A superior Russian officer once told me, that when he first left his native country on a visit to the west of Europe, he did not know what to do with the sheets which he found in his hotel bed, and that he slept outside of, instead of between them! This circumstance may give some idea of the comforts which the middle classes of Russia are likely to expect at hotels, and though, perhaps, it seems a monstrous thing to say, it is a fact, that people who don't use sheets on their beds, may be most undeniably good officers.

# CHAPTER II.

## SAMARA TO ORSK—*continued.*

The steppes east of the Volga—Gloominess in winter—Their bareness of timber—Inconveniences of post route—Streams and rotten bridges—Delays at ferries—Peasant superstitions—Mud in villages—Post-house—Travellers met—Imperial portraits—Image of saint—Village churches—Piety and superstition—Saints' days—Vodka-drinking—National intemperance—Hand-swipe for raising water—Bird-cots—Debateable ground between Europe and Asia.

As I travelled into Asia in the early spring, while my return journey took place as winter was setting in, it is probable that my impressions of the great tracts east of the Volga may partake of the gloom which the landscape usually presented on both of these occasions. The first journey was naturally the more cheerful of the two, for bright green young grass sprouted through the remains of snow-drifts which were still seen on the steppes near Orenburg, and further east, about Orsk, the pasturages were covered with a thousand gay wild flowers. But mud was the chief feature of the country, both in May and October, while in the latter month frost, snow, and cutting north-east winds were the further fitting accompaniments to the desolation of the salt deserts near Lake Aral, and to the monotony of almost endless plains near the Volga. The dreariness of the route is heightened by the scarcity of timber, and the blanched stems of the birch-trees, which are sometimes seen, add to, rather than lessen, the mournful aspect of the flat, far-stretching country. The quantity of wood used for building and for fuel is enormous, and the effect of these conditions is evident on the bare steppes, deeply furrowed by ravines which are excavated by the rapidly-escaping drainage; a serious evil, which the Government, as well as many private landowners, have commenced to combat by plantations, though the areas to be dealt with are so extensive that a passing traveller can scarcely recognise such action as has already been taken in this direction. About the Volga, the Ethnic affinities of the population with the Turanian races have possibly assisted this leafy denudation of the neighbouring districts, and the spirit of the old proverb, that 'he is no true Turk who reposes in the shadow of a tree,' is in entire unison with the nakedness of the country. Gardens are such a rarity as to convince one that a prejudice must exist against them; and though trees of moderate size are to be seen in the valleys of the Samara, Sakmara and Ural rivers, any owing their existence to artificial means may probably be searched for in vain.

At a time of the year when streams are swollen and rivers are in flood, the post track, which runs across country, and usually parallel to the main watercourse line, is naturally not free from many difficulties. Numerous secondary streams have to be crossed, from whose deep muddy beds nothing but a vigorous application of the *yamshik's* whip to the team of straining horses, can extricate the groaning *tarantasse* of the traveller. Ricketty wooden bridges exist over some of these drainage channels, but the long and severe winter rarely leaves such frail structures in a serviceable state at the opening of spring. Usually the *yamshik* will trust to the strength of the carriage, the toughness of the harness, and the goodness of the horses, and prefer to take his chance in the mudholes, rather than go through that service of danger which is implied in crossing a rotting timber bridge. If, by bad luck, the vehicle should stick in the bed of the stream, there are generally a sufficient number of good-natured peasants passing, who will willingly lend a hand to drag it out again to *terra firma*. Later in the year the bridges are put in order, though the local authorities are evidently in no hurry about the work, for I noticed that most of such repairs had only just been completed on my return journey in October.

Sometimes the traveller will be delayed at a larger rivulet, which the ferrymen decline to cross over until the flood shall have subsided; and on one such occasion I passed nearly twenty-four weary hours, in company with many foot passengers, on the banks of a broad swollen stream, whose passage was made dangerous for the ferry-boat by the strong wind which was blowing at the time. My fellow-sufferers were still more impatient at the delay than myself, and, amid much unreasonable grumbling, attributed our misfortune to the presence of an unhappy and almost guilty-looking priest, who also wanted to cross over. On enquiry I learnt that Russian peasants have curious superstitions regarding the village 'pope' or priest, considering it unlucky, for example, to meet him when they are on the point of starting on any excursion or journey. Whenever, consequently, the great men of a village happen to be going out shooting, the 'pope' makes a point of keeping quietly at home, in order to avoid the possibility of any unlucky *rencontre*.

The mud of the post route across the open country comes to a climax in the villages, which are passed through at about every fifteen miles between Samara and Orsk. In these, two lines of weather-beaten wooden shanties form the bounds of a deep broad sea of black slush, trodden knee-deep by passengers, or by flocks of cattle, which come trooping out of farmyards, in the grey morning dawn, on their way to the surrounding pasturages. The *tarantasses* must afford a welcome break to the dulness of these far-away villages; and as the steaming, jingling teams go galloping through them, peasant women, with labour-lined faces, gaze with a somewhat languid curiosity through the small windows of the log-huts, while white-haired village urchins stop their play for a moment, and the pretty little girls sitting on the door-steps glance shyly up from their knitting. It is scarcely any wonder if these children, who are literally *selected* lives, should look strong and healthy, since two-thirds of the offspring of Russian peasants are said to be killed by hard conditions of life before they reach the age of twelve months!

Passing by the men, who, wrapped in their sheepskin pelisses, scarcely deign to notice the disturbance, and tramp doggedly through the mud, intent upon their affairs of the moment, the traveller's *tarantasse* will reach, somewhere in the village, an open space, on one side of which is generally the post-house. Handing over his order for fresh horses to the post-master, who will record his passage in the station-book, the traveller stretches his cramped limbs and calls for a *samovar* to refresh himself with a cup of tea, while the new team is being harnessed and put to. Sometimes, and more especially near Orsk, the post-master's neatly-dressed wife or daughters (whose small feet are shod with *bottines* which are noticeable even among the capital boots made in Russia), will bring out soft and beautifully-netted woollen shawls for sale. These are of all sizes and prices, a very good one, about fifteen feet square, being worth as much as thirty pounds sterling; they are excedingly light and warm, and are in great demand among fashionable Russian ladies as wraps for winter. Hard bargaining takes place over their purchase; and the keenness of the women will be found to justify one of Napoleon's favourite sayings, that it 'takes three Jews to deceive a Russian.'

Between the Volga and Orsk there is a considerable amount of movement along the post-route, and land proprietors, farmers and peasants are met in numbers on their way to or from the village fairs, while at the post-houses there is in general no lack of passing travellers, who are ready to enter into conversation and to exchange notes of the way. These travellers are for the most part military and civil officers, besides a few merchants, and they are not unfrequently accompanied by their families; the number of babies in arms, indeed, who are met with, and whose affairs seem to take them a post journey of 2,000 miles into Turkestan, is perfectly astonishing, and says volumes for the *physique* of Russian mothers. In addition to the above, native traders from Central Asia will now and again be seen, either taking their tea, or going through their devotions, on a praying carpet, in the corner of the travellers' common room.

The walls of the common room of the post-house are adorned with coloured lithographic portraits of the Emperor and chief members of the Imperial family, whose features are so brought to the knowledge of the many races that people the empire. On seeing a half-savage Kirghiz gazing intently at the portrait of the Czar, the idea could but occur whether the English might not take a hint from the Russian example? During many years passed in India I saw but one portrait of Her Majesty in a public building, and that one was not accessible to the common people; under such circumstances, it can scarcely be matter of astonishment that our Indian fellow-subjects, who are nothing if they are not influenced by outward and visible signs of authority, should be indifferent to that mysterious abstraction which is known to them as the Government of India.[1] *Apropos* of the above remarks, it is a curious circumstance that the features of the two sister princesses who will eventually share the thrones of England and of Russia could be made more easily known to a greater number of Asiatics than of Europeans!

The postal contractors supply the station-houses with the Imperial portraits, as well as with the

[1] Time will show the nation what it owes the Prince of Wales for his recent happy inspiration.

framed and gilt image of the saint, which is an indispensable article of furniture, even in the poorest Russian hut. The feeling of repugnance which may sometimes be evoked by the superstition of the followers of the orthodox Greek faith is considerably modified in passing from the great centres of credulity to the quiet villages to the east of the Volga. Close to the post station is the village schoolhouse, and, in the centre of the open square, stands the church, with clean whitewashed walls, which are panelled in coloured pictures of saints and martyrs. Crowned with a great green dome, the church is made still more conspicuous by the glittering metal spire of its campanile, which towers hard by; and though the architecture is *bizarre*, and the ornamentation is meretricious and gaudy, the broad fact remains that the best building in a Russian village is the house of prayer. It is only in communities having a higher development that cathedrals and stockjobber's palaces stand in open rivalry, or (as at Moscow) that the carriage-wheels of Phryne and Laïs splash the equipage which takes the picture of the Mother of God for an airing. But upon these distant Eastern steppes one of the most frequent sights presented to the traveller is a band of peasants, mostly aged men and women, who, with staff and scrip, are on a weary pilgrimage to some famous monastery of 'Holy Russia.'

Saints' days, according to the calendar, are excessively numerous in Russia, in spite of the fact that the present Emperor has considerably diminished their number by a special ukase. Among the ceremonies which take place upon these occasions, hard *vodka*-drinking fills no inconsiderable part. I did not see many drunken men on my journey across the steppes, but, as a compensation, I remarked numbers of women, who were at least excessively gay, and dancing and singing in the various villages passed through. The men were probably much more tipsy, and stowed quietly away somewhere—an inference which is not so unjust a one as it may seem at first sight, for the Russian peasant, when he does drink, swallows tumbler after tumbler of vodka, until he falls senseless, and is carried home to sleep off the liquor.

Nor must it be thought that the vice of intemperance is confined to the Christian population of the villages between the Volga and Orsk, where the Muhammedan element increases as one travels to the east, and where the slender minaret of the mosque not seldom stands at no great distance from the dome of the Greek church. If the Muhammedans were sober, and the Christians drunk, during the feast of the Pentecost, which occurred during my journey into Turkestan, the nights of the Ramazan, on my return, were made hideous by the debauchery of the Moslems ; and I was told the Christians availed themselves of Muhammedan hospitality on this occasion, and so doubled their opportunities of slaking an inordinate thirst for *vodka.* The cheapness of this spirit is a terrible incentive to intemperance, and the Government are understood to be considering measures for mitigating the effects of the national failing. In this respect, the recently-introduced law of obligatory military service will act somewhat as an educational measure, but the general advancement of culture, and an increase to the means of procuring rational recreation, are the only conditions which can really arrest the existing proneness to this vice in Russia.

In all villages east of the Volga are two common objects, which indicate the mixture of European and Asiatic population, on the limits of the West and East of antiquity. One is the machine, which has been in use for ages in every Oriental country, for raising water from wells—the hand swipe, consisting of a bucket, suspended from the end of a long fixed lever; the other is the little birdcot, fixed upon the top of a high pole, close to every peasant's hut. This latter object I noticed along a continuous line of more than 3,000 miles, from the shores of Lake Constance in Bavaria, to the banks of the Syrdarya in Turkestan, and, I was told, it can be traced much farther still to the east.

If Nature has been far from prodigal in gifts of beauty to the debateable lands lying between Europe and Asia, the deficiency is amply compensated by the crowd of past memories which environ them. Across this gloomy country, which led to the haunts of the gold-guarding griffins, passed the track of the early Turanian tribes as they advanced to the occupation of the countries of Europe; and here Darius, the son of Hystaspes, marched with his perplexed Persians in pursuit of the ever-retreating Scyths. The ancient Greek caravans, which left the marts on the shores of the Euxine to bring back Siberian furs and minister to the luxury of semi-mythic nations, travelled over these plains; where, in later times, the envoys of popes, emperors, and kings of the West passed with messages of peace to the great Asiatic conquerors, whose devastating armies (it was feared) might descend upon Christendom. Over these regions surged the advancing and retiring waves of the Golden Horde of Kipchak, while in our own days a reflex tide has set in towards the East, to bear with it the germs of events which may even surpass in importance the long chain of past deed and thought attaching to this antique route of humanity.

# CHAPTER III.

## ORSK TO IRGEEZ.

Russian indifference to time—*Tarantasse* journeys—Hardships of early days of travel into Turkestan—Improvements of present day—Government couriers and Kirghiz jiggits—The Bashkirs—The Kirghiz steppe—Country from Orsk to Karaboutak and Irgeez—Its possible physical amelioration—Advantages to Russia of the nomadic state of the Kirghiz in these localities in particular, and generally in West Turkestan.

THOUGH the distance of one thousand miles which separates the Volga from the Syrdarya is not a very disagreeable journey to make in a comfortable *tarantasse* with post-horses, probably few but Russians ever accomplished it without regretting their hard fate. The idlest man of Western Europe would at least deplore the loss of time which is involved in such a trip; but this is a notion which is scarcely comprehensible to the mind of an average Russian. Railway travelling may eventually remedy the defect, but meanwhile, Russians are, in general, entirely ignorant of the value of time, and it was not till I went to Russia that I could explain why a gentleman of that nationality, whom I had met in Italy, used to take frequent opportunities of repeating the old English saying 'Time is money,' with a *gusto*, as if he had made some great and original discovery. A colonel of my acquaintance, having had occasion to visit his general at head-quarters, missed the only train he could go by on three days in succession, and his wife, from whom I heard the story, told me he consoled himself for his dilatoriness by remarking that such disagreeables were impossible before the railway was open! Any number of anecdotes on this head could be related, and all classes in Russia are saturated with this remnant of barbarism; even the business men are complained of by English merchants for their utter indifference to punctuality. 'It is all very well,' said one such merchant to me, 'to talk about the patience of Job, but Job never lived in Russia.'

If a Russian pays no heed to the time which is wasted during a long *tarantasse* journey, still less is he sensible of the corporeal inconveniences of this mode of travel; for, feeding on occasional cups of tea and continual cigarettes, he pursues his wearisome journey across the steppes, a very Buddha of contemplative inaction, and totally unconcerned, to all appearance, with the past, the present, or the future. I once expressed my astonishment that even women seemed to find no inconvenience in travelling such great distances, and was told, in reply, of a lady who went yearly from Poland to Irkutsk to make a visit to her mother, a journey, counting both ways, of something like eight thousand miles! Russians, in short, both men and women, think no more of travelling this enormous distance with post-horses than Englishmen and their wives and daughters think of going to India in a P. and O. steamer.

Even in the earliest days of Russian advance into Asia, *tarantasse*-travelling as far as Orsk may have been a comparatively easy matter, since villages and a settled population were not wanting, and the physical aspects of the country were favourable. But before the service of post-horses was organised farther to the east, grave inconveniences and often hardships were experienced by Government *employés* and their families travelling into Turkestan. Provisions were hardly ever procurable, even from the few scattered *aouls*[1] of the Kirghiz; and water, that first necessary

[1] Nomadic villages.

of life, was not seldom entirely wanting. Delicate women were rendered half frantic by the death of their children from insufficient or unwholesome food; and sorely-tried strong men sought a solace for their troubles in the bottle. Report says that, at this epoch, months were passed upon the road from Orsk to Cazalinsk on the Syrdarya, a journey which is now a matter of days. The post-stations, which have been built along the route crossing these desolate regions, afford excellent accommodation for travellers, and wells have been dug along the whole distance, though it is true that the water in many of them is not of good quality. If the traveller to Turkestan will do well to carry a small store of provisions with him, he will certainly not starve at the present time should he be unprovided.

The good old times are fast passing away, for the iron horse will, after a few short months, be heard snorting its way between Samara and Orenburg; and the couriers, who formerly left St. Petersburg at a moment's notice, and travelled in an open post-waggon thousands of miles across Asia, lament, it is said, the degeneracy of these days of railways. Even the Kirghiz *jiggits*, who rode with a despatch from Tashkend to Orenburg, and returned with an answer within twenty-four days, have been now superseded by the telegraph wire. Such a courier, riding one horse and leading another, which he would mount when the first became tired, used to get over the ground at about one hundred and thirty miles daily, and required but a few hours' rest at Orenburg before starting again on his return journey. My Lord Chesterfield himself would scarcely have regarded such a feat with the disdain he apparently entertained for the performances of post-boys in general.

While the population in the valley of the Ural, between Orenburg and Orsk, is in course of being throughly Russianised, the country north of the latter place is occupied by the Bashkirs, a race of Turanian origin, who, though now semi-sedentary in their habits, were formerly nomadic shepherds. In the summer, they still take their herds of cattle to the pasturages on the wooded eastern slopes of the Ural chain, but in winter, they return westward to villages which are situated in the mountains, and which they built, according to the Government order, some years ago. At first they strongly objected to live in these villages, so that it used to be pleasantly said that a Bashkir village was the cleanest village on earth—as it very well might be, seeing that no one but an old man, specially told off to look after the houses, ever inhabited it. But sedentary habits have gradually gained ground, and with their growth, the government of the Bashkirs, though still administered by the chiefs of the tribe, has been closely assimilated to that of the rest of Russia. Civilisation steadily makes its way, and many young men of the richer Bashkir families receive a university education, while not a few are officers in the army. It is probably a mere question of time when agriculture and mining industry will take the place of pasturing sheep and cattle among these people; and their complete reclamation from the pastoral state will afford encouragement for that of their neighbours the Kirghiz, who a century hence may perhaps hold the same place in the scale of Russian civilisation as the Bashkirs do to-day.

The great Kirghiz steppe is the eastern portion of that belt of low-lying country which stretches from Europe into Asia, along the frontier of Southern Siberia, and is divided by the upheaved chain of the Ural mountains, which run in a north and south line. In a south-westerly direction these elevations are prolonged by the hills which are situated on the left bank of the Ural river, and whose slopes between Orsk and Orenburg fall towards the Caspian ; while in a south-easterly direction the Ural mountains fall towards the basin of Lake Aral. They are prolonged directly to the south, under the name of the Mougodjar hills, which terminate about twenty miles from the elevated Ust Urt plateau, situated between the two seas. The Kirghiz hordes occupy the low plains which, commencing from the north-eastern shores of the Caspian, and continuing along the Emba steppe, passes across the country north of Lake Aral, directly to the eastern end of Balkash. They also inhabit the banks of the middle and lower courses of the Syrdarya as well as the Kizzelkoom deserts, which are localities taken possession of by them in more recent times.

It is upon the tract of country, five hundred miles long, separating Orsk from the lower Syrdarya, that the physical characteristics of the ground have placed such great difficulties in the way of the Russian advance in the past, and still render Russia's line of communication with Central Asia a highly inconvenient one. Here are no villages nor fixed population, though two small outposts have been established, to contain the Government establishments required for the administration of the nomads who in spring and summer wander northwards, to the pasturages on the slopes of the Mougodjars and south-eastern Ural. The more westward of the two posts is Fort Karaboutak, which is situated at one hundred and fifty miles from Orsk, upon the edge of the high ground sloping down to the northern portion of the basin of Lake Aral. It is on a cliff which overhangs a stream of the same name, one of the affluents of the Irgeez, and is surrounded by about one hundred wooden houses, tenanted by families of Cossack settlers. From Orsk it is reached in a south-easterly direction by a gentle ascent, leading up to the plateau on the Mougodjars, and though the intervening distance is entirely bare of timber, it is not wanting in good pasturages, where a thousand bright-hued flowers are seen in spring, amid tolerably luxuriant grass. Nearly the whole of this country is well watered by small streams, running to the Ural river or to the Irgeez, and flowing sometimes in ravines, which are formed by bold purple-tinted masses of porphyry, or of weathered granite, or of green stone. From Karaboutak eastwards, a gloomy sterility becomes the leading feature of the scene, and the grassy lands which have just been passed over are replaced by stony or dreary bare tracts of clay as the slopes of the Mougodjars are descended. Water gets scarcer in the small affluents of the Irgeez, while such herbage as grows is almost too bitter even for camels to thrive upon. The river Irgeez, which flows south-east, to lose itself in salt quagmires and lakes, changes, as the short spring passes by and the hot summer sets in, from a wide shallow stream to a chain of stagnant salt pools, which are separated by sandbanks. On a low swelling knoll upon its banks, about one hundred miles from Karaboutak, is situated the second outpost, called after the river, Irgeez. The inhabitants of this place would be condemned to drink salt water during nine months of the year, if they had not adopted the plan of damming the flood stream in the spring, and so diverting the freshly-melted snow-water into artificial store reservoirs. It is barely perhaps a question whether, by an extension of this device, much water which now runs off and is wasted in the salt tract lying to the north of Lake Aral might not be utilised in ameliorating the sterility of the country and supplying it with drinkable water.

This, however, could only be done as far as Irgeez, for east of this place commences a tract of salt deserts, which will be presently described, and will be seen to be incapable of improvement. Admitting, however, the possibility of improving the country west of Irgeez, what has been said is sufficient to show that years must elapse and an enormous capital be expended before it could be made fit for permanent occupation; and it therefore seems to result that the Kirghiz visiting these localities should be left in the pastoral state rather than be improved into sedentary and poor agriculturists.

Until the present line of communication with Asia be replaced by a railway—a project which will scarcely bear any examination, from the broadest financial view, for many years to come, even if it ever does so—it is a necessity for Russia that the Kirghiz frequenting these localities shall retain their nomadic habits. Among these people, more perhaps than among other Asiatic tribes, the circumstances which necessitate frequent moves across enormous distances have been intensified by the particular conditions of the steppe they occupy. Brigandage, which Russia has by immense pains and sacrifices eradicated amongst them, was the natural employment of these shepherds, who wandered over the great Asiatic plains traversed by rich trading caravans. The old inter-tribal jealousies and the frequent blood feuds have also lent a charm to these broad lands, across which a distant flight would ensure safety from a pursuing enemy, or a haven from the persecution of a stronger race. This love for a wandering life causes, at the present day, a continual small stream of passing population to oscillate across the Aralian steppes; and such movement modifies, in a very sensible degree, the inconveniences and the dreariness of the present line of communication with Turkestan.

It is perhaps no exaggeration to say that, had the aid thus afforded by the Kirghiz been wanting, the Russians would scarcely have succeeded, up to the present day, in making their way into the valley of Syr, which has formed a broad and easy highway into Central Asia. Knowing every inch of the intervening wastes, and possessing vast numbers of camels, sheep and cattle, these hardy people provided the guides, the transport and the commissariat which were required for the Russian advance into Turkestan. What the services they have thus rendered have been may be judged of in an indirect, though very significant, manner, by considering that since the misfortune which occurred to Peroffsky's army in 1839–40 (when relations with the Kirghiz were scarcely on the friendly footing which they are on to-day), no hitch or hindrance whatever has been experienced by the Russian forces in their operations in Asia. There are few, if any, instances in the military annals of any European nation where the generals have been more spared those labours and anxieties which attend expeditions in barbarous and desert countries, and which arise from difficulties of transport and supply. Whether the assistance afforded has not encouraged the national Russian tendency to unmethod, is however a question which is suggested by at least two episodes of the recent Khivan campaign, where disaster seems to have been averted from the Russian arms only by the fortuitous discovery of water by the Kirghiz at Alti-Kuduk, in the Khalatta desert, and on the Ust Urt plateau, between the Caspian and Aral Seas.

# CHAPTER IV.

## THE ARALIAN STEPPE.

The depressed tract round Lake Aral—Its separation from the Valley of the Obi—The overflow of its basin—The pre-historic Mediterranean, which included the Black, the Caspian, and Aral Seas—Deserts north of Aral—Postal arrangements across them—Horses and camels—Traffic *en route*—Kirghiz on the move—Their regulated wanderings—The *arundo* of the Aralo-Caspian region—Its importance on local physical phenomena.

ALTHOUGH European civilisation may be said to cease at Orsk, and the Asiatic steppes are there entered upon, no break exists in the postal communication which Russia has established with Turkestan. The more habitable portion of this distance has already been described, and it remains to submit to my reader a sketch of the physical aspects of the deserts situated in the basin of Lake Aral, so that the difficulties of this part of the route, as well as the spirit with which those difficulties are met, may be judged of.

The gap which separates the Mougodjar hills

from the Ust Urt plateau, and in which the Tchagan stream has its commencement, connects the low sandy steppes, crossed by the river Emba on its way to the Caspian, with the depressed tract of country which lies north of the water-spread of Lake Aral. This depression would be continuous, like a furrow, along the eastern foot of the Ural chain as far as the Frozen-Ocean, but for a low transverse ridge which crosses it in an east and west direction in latitude 51°, and so separates the basin of Lake Aral, on the south, from the sources of the Tobol and Ubagan rivers, which are affluents of the Obi flowing to the north. The surface of the country about this transverse ridge is covered with a number of small and ever-decreasing salt lakes, which have been supposed to be the remains of a great bitter sea placed by ancient Chinese legends somewhere in this part of Siberia; but which, as will be presently seen, there are very good grounds for ascribing to the overflow in a northerly direction of the salt lakes which are situated to the north of the present water-spread of Aral, at a time when they were united in a single body of water whose surface possessed a much higher level than any one of them has to-day.

It is almost certain that an overflow of water

took place from the basin of Lake Aral, in a westerly direction through the gap lying to the south of the Mougodjar hills and flooded the steppes between that neighbourhood and the Caspian. With a level allowing of such an escape, the waters of Lake Aral would also have passed through at least one cleft in the cliff, which is at once the eastern limit of Ust Urt and the western shore of Aral; and would have filled up some of the numerous salt lakes and depressions covered with sand, which are found on the surface of this highland. At the south-western corner of the basin the surplus waters would have found their way over the surrounding edge and have inundated the tract of country now traversed by the dry Uzboy bed of the Oxus, on its way down to the Caspian Sea.

Such seem to have been the aspects of these regions in historical times, and as regards their pre-historical aspects, it seems tolerably certain that, previous to the opening of the channel of the Bosphorus, an immense freshwater inland sea included in its water-spread the basins of the Euxine, the Caspian, and the Aral, and discharged its surplus waters into the valley of the Tobol over the low ridge of ground which transversely crosses the depressed furrow extending in the direction of the Northern Ocean.

Omitting, for the present, any consideration of the relative heights of different localities in the Aralo-Caspian depression, a very striking indication of a former existing state of things is afforded by the labours of natural historians. A Russian naturalist, M. Bogdanoff, has observed that the *fauna* characterising the Polar Sea are found in continually diminishing numbers, as the salt lakes extending from the north towards Lake Aral are successively passed by. It has been established that the *fauna* of the Aral, the Caspian, and the Euxine basins are almost identical, while the researches of Fuchs and others have shown that those of the last-named sea differ in comprising species which are characteristic of the Mediterranean; a change which is more specially notable near the outlet of the Bosphorus.

In travelling therefore from Europe into Asia, the part of the route immediately after Irgeez crosses a depressed country, which is the bed of an ancient sea, and in this hollow, now occupied by masses of sand of greater or less depth, are also salt lakes and quagmires fed by the Irgeez and Tourgai rivers, and by other streams flowing down in spring from the northern limits of the basin of Lake Aral. Ground capable of cultivation, it need scarcely be said, is not to be found upon these sandy, salt, clay tracts, where numerous beds of gypsum and selenite occur in combination with layers of marine shells. The masses of sand are extensive on the north-west of the water-spread of the Aral, but assume their greatest dimensions in the Karakoom (black sand) deserts on the north-east. Anything more desolate and hopeless-looking than this country would be difficult to imagine, for along many scores of miles a few stunted artemisia bushes are the sole vegetation to be met with. In the Karakoom there is perhaps a change for the better, for there the undulations are clothed with larger bushes, whose dark colour presents a not unpicturesque contrast with the lighter colour of the sand-hills they stand upon. Animal life in these deserts is of course scarce, though in the shallow lagoons there are numbers of wild-fowl, including all varieties of storks and cormorants, as well as of wild ducks and geese. Insect and reptile life is more plentiful, and at many post-stations it is amusing to watch the beetles (*Ateuchus sacer* ?), to which allusion is made by Pliny, who describes them as walking backwards and rolling balls of dung, in which they deposit, and so guarantee against the rigours of winter, the *larvæ* to perpetuate their race. The stronger and more cunning of these *scarabæi* leave to the weaker and more simple ones the labour of rolling the small balls of camels' dung to their holes from a distance, when they rush out from an ambush and carry off the coveted spoil.

Across the desolate Aralian deserts, which are almost entirely deprived of vegetable and the higher forms of animal life, post-stations have been constructed, and wells, whose contents are more or less salt, have been dug at every twelve or fifteen miles' distance. Usually an ample supply of horses is found, and little delay occurs in harnessing a fresh team, though sometimes a deficiency of drinkable water will require the detention of the traveller at some station until the animals return from a well containing some of better quality in the neighbourhood. The postal arrangements, which, considering all the circumstances, are as near perfection as they can be, are made by the Government, for a fixed term, generally of from five to ten years; and this is rendered desirable, if not necessary, from the contractor having to provide everything on the route, commencing with the repair (and even the construction) of the post-station and subsidiary buildings, such as stables, &c., and ending with the nails for the horses' shoes ; that is, on those portions of the route where the stony nature of the ground demands the shoeing of the horses.

East of Orsk, the owners of the post-horses are generally Kirghiz, who procure their cattle from relatives, established either on the lower Volga or Ural or from Siberia ; for the animals from the Syrdarya, though cheap and numerous, are smaller and inferior in quality. In Siberia, a post-horse, which usually stands about fourteen hands, costs about 3*l*., a sum which is considerably less than its value upon the Aral steppe. Such animals are well fitted for postal purposes, and generally accomplish the whole distance of a stage at a hand gallop ; they are excessively hardy, as indeed they ought to be, when the severe work they do and the scanty fare they get are considered. From Orsk grain has to be carried eastwards, while forage is scarcely to be had at all, except about the plateau on the Mougodjars in the early spring, and, in the more desert parts of the route, the animals are often left, for twenty-four hours or more, without a feed at all ! Probably indeed, in no other part of the world, could conditions so favourable be found for the untrammelled action of the principle of the 'survival of the fittest.'

At one or two stations, where forage is less scarce in the spring of the year, a large number of horses are sometimes collected, and I was told, at one such place, that there were more than one thousand animals upon the pastures of the steppe in the neighbourhood. Such a number represented a capital of about 5,000*l.*, and the fact affords some notion of the benefits which have accrued to the Kirghiz from the advance of Russia into Asia, for a century ago the same horses at such a place, would scarcely have been worth more than as many shillings.

Nomadic as they are in predilection, not a few individual Kirghiz find employment along the post route as *yamshiks* and grooms, and live in a mud-roofed barrack or a *kibitka*[1] or two, which are pitched in the neighbourhood; so forming, with the small Cossack guard of the place, a little local population. Russians, as is well known, are excellent coachmen, and the Tartars near Orenburg and Orsk and the Kirghiz of the steppe, are scarcely

[1] Felt tent.

their inferiors in the art of driving a team. The traveller from the west of Europe is, however, not unlikely to class them, in the order they have been named, as being brutal, more brutal, and most brutal, in the treatment of their horses.

It is not in these animals alone that the Kirghiz reap advantages from the postal traffic into Turkestan, for at some few stages in the Karakoom desert draught camels are supplied to draw the traveller's *tarantasse.* A team of camels is of course slower than one of horses, even across heavy sand, but they get less distressed; nor do they require as much water, nor of so good a quality, which are considerations of importance upon the great sandy waterless tracts skirting the edge of Lake Aral. Excluding sick or reserve animals, the daily requirements at each postal-station east of Orsk appear to be six teams, or say twenty horses; though more to the west the number would be much larger. In addition to the post-waggon, and allowing three souls for every team, we arrive at fifteen travellers daily, as the number posting out of Europe into Asia, and *vice versâ.*

Independently of this traffic, east of Orsk will be seen many carts and camels laden with hides and cotton from Central Asia, or carrying back iron and manufactured goods in return ; and of Asiatic produce I estimated I saw about thirty-five tons of cotton and twenty tons of hides, during two days' journey, at the height of the transporting season. Once at least, on the way, the traveller will pass a *sotnia* of Cossacks or a detachment of infantry, on their long weary march into Turkestan; and, in the early spring, the solitude on the Mougodjars is enlivened by herds of grazing camels, whose numbers, in one case, amounted to at least six hundred, and consisted chiefly of dams with their foals.

Not seldom also will caravans of Kirghiz be encountered, every man, woman, and child of whom are mounted on horse or camel back, and who, conveying with them their *kibitkas*, household goods and property of every sort, drive their flocks and herds in front of the cavalcade in search of fresh pasturages and water. Such wanderings of the nomads are methodical, and their movements take place in a known and predetermined circle of change; so that any man of a certain family, of a specified *aoul*, belonging to a particular tribe of Kirghiz, can generally be found by the Russian authorities when he is wanted.

These people still retain their own special laws,

which are administered under the patriarchal government of their chiefs, though they are also subject to the authority of the Russian head of the district in which they may happen to be wandering at the time. From the eastern shores of the Caspian, and from those of the Aral, the Kirghiz travel northwards during the spring and summer months, near the sources of the Tourgai and of the Tobol; but as winter sets in they return again southwards, where broad tracts of low-lying country covered with high rushes (*arundo*) afford them fuel as well as shelter against the rigours of the climate of that season.[1]

The whole of the sandy margins of the lakes and marshes, both of fresh and salt water, as well as the banks of the rivers of the Aralo-Caspian country, are covered with a luxuriant growth of these rushes, which are used as cattle fodder, besides affording to the nomadic population the advantages which have been mentioned; and the existence of this aquatic vegetation, however natural it may appear to be, is deserving of more notice than has hitherto been accorded to it, and has had a very important and special bearing on the physical aspects of these

[1] The geographer Strabo mentions these as being ancient Scythian habits.

regions. For example, upon the issue and consequent lateral diffusion of the overflowing waters of Lake Aral over the moist bed of the pre-historic sea which formerly existed in this part of Asia, the stems of these plants, which grow closely together and attain a height of between fifteen and twenty feet, would have prevented the escaping waters, on their passage to the lower level of the Caspian, from acquiring the momentum necessary for the excavation of a regular and pronounced river-bed, into which such an overflow resolves itself in all ordinary cases. To this apparently insignificant circumstance may in all probability be ascribed the absence of those physical traces of overflow whose recognition would have placed the theories of physical geographers regarding Lake Aral more in accord with the actual facts of its past history.

# CHAPTER V.

## EARLY RUSSO-ASIAN POLICY.

First commerce between Russia and Central Asia—Englishmen engaged in this trade—The brigandage of the Kirghiz hordes—Necessity of extirpating it, and of preserving frontier tribes under domination of Russia—Flight of the Volga Calmuks—Submission of Kirghiz—*Divide et impera* policy with frontier tribes—The Bashkir rebellion—The Kirghiz opposed to this people—Eventual subjugation of Kirghiz and pacification of Asiatic frontier.

THOUGH something more than six centuries have elapsed since Russia first had commercial relations with the Central Asian Khanates, by way of the Kirghiz steppe, it was not till nearly the end of the fifteenth century that a systematic attempt was made to develop her trade in this direction. During the sixteenth century we find Englishmen engaged upon the project, and many commercial agents, of whom Anthony Jenkinson is a good type, pushed their way into these regions. Some time subsequently an English naval captain, named Elton, was to have accompanied a State caravan into Central Asia and to have reconnoitred Lake Aral, with a view to the establishment of a naval flotilla upon it. In 1742, Gladychef, the Russian envoy to the Karakalpaks of the Lower Syrdarya, met an English merchant, Djaké (? Jacky) who was trading with the Kirghiz and of whom he bought various articles as presents for the tribe to which he was accredited. And in 1745 we find Jonas Hanway, 'the handsome Englishman,' trading and travelling in these countries before entering upon his career of usefulness and philanthropy at home. Looking back upon these few and incomplete facts, it cannot be doubted that Englishmen formerly filled no unimportant part as pioneers of commerce and of civilisation in Central Asia, a state of matters which Humboldt has noticed and which perhaps afforded M. Vámbéry the grounds for his remark regarding the apathy of the Englishman of the present day.

The Russians had some dealings, through the Nogai Tartars, with the Kirghiz in these early times, and many unsuccessful attempts were made to enter into closer relations with them; though it was not till the Siberian conquest that the two nations were brought into direct communication through the Khan of that territory, who was himself a Kirghiz.

At the end of the sixteenth century, the first proposition was made to Russia for the reception of these people as subjects, but the negotiation fell through, and acts of hostility and brigandage continued to be the normal conditions of the Kirghiz steppe until 1730, when these nomads became subject to the Empress Anne. In 1739 the first great commercial caravan was despatched from Russia to Tashkend, and this was plundered by the Kirghiz of the Middle Horde, at two days' march from its destination. With perfect impartiality the Kirghiz of the Little Horde, a few years after, plundered a large Khivan caravan on its return from Russia; and, desirous as both Russians and the inhabitants of the Central Asian Khanates were for an advantageous interchange of products, the cause of progress was hopeless while these acts of brigandage continued.

The task which devolved upon Russia was not merely the extirpation of this evil on the Asiatic steppes, but the conciliation at the same time of the Kirghiz, for however possible it might have been to drive these people to the east by a succession of raids made upon their pasturages, such proceedings would not have advanced the Russian object of placing her commercial relations with the Khanates upon a secure basis. It was not only a cessation of brigandage which was required, but the existence of some sort of population, more or less frequenting the intermediate country, and serving as links of communication while lessening the inconveniences of this trade route. Russia's aim was therefore to keep the Kirghiz under her domination, and the tenacity with which she has always followed this idea in dealing with the populations of her Asiatic frontier is well exemplified in the episode of the flight of the Volga Calmuks to their old eastern homes, on the frontier of China. This little-known but highly-interesting event took place in 1770, and a short *résumé* of it may be inserted in this place.

The original country of these Calmuks, or more properly Torgoutes, was in the west of China, where being oppressed by the continually-increasing power of the Zungars, the great mass of Calmuks emigrated in 1636 to Russia, and were granted pasturages on both banks of the Lower Volga by the Czar Michael Feodorovitch. After the destruction of the Zungar power by the Chinese in 1756, the remnants of the Calmuks remembered and rejoined their compatriots in Russia, and brought with them, as might be expected, entirely different ideas to those held by the larger section of their tribe, who had lived peacefully for more than a century in the west. These were habituated to the ease of a quiet life, and those to the alarms and agitations of war; yet, strange to say, the latter minority succeeded in arousing in the breasts of the former majority an ardent desire of returning to their old nomadic life in China.

A great portion of the Calmuks being established on the right bank of the Volga, the freezing of the river was the date chosen for the move. It turned out, however, that the winter of 1769–70 was a mild one and the river did not freeze, which is the reason why some eighty or one hundred thousand Calmuks, of the Buddhist religion, are found in the government of Astrakhan at the present day. The greater portion of the tribe, to the number of 30,000 *kibitkas*, or 150,000 souls, set out, however, from the Lower Volga, on January 5, 1770, on their march to China.

As they approached the river Yaïk (the Ural of to-day), the governor of Orenburg, with a view to preventing this emigration, informed the Kirghiz of the Little Horde that the Calmuks were advancing to attack them, and the Siberian government gave similar warnings to the Middle Horde and to the Great Horde, who lived further to the east of the steppe. Animated by hate and lust of booty, all the tribes of the Kirghiz, from the Yaïk to the frontiers of China, assembled in armed bands to dispute the passage of the ill-starred fugitives, encumbered, as these were, with their families, cattle, and household goods.

The Cossacks of the Ural first opposed the march of the Calmuks, who speedily broke through, and were troubled no more in this locality. The Orenburg Cossacks were then sent to reinforce the Little Horde Kirghiz, but they accomplished nothing, from their horses being knocked up; and though regular troops were also sent forward, the Calmuks had already passed the Aralian steppes, and the Russians, who had suffered greatly from want of provisions and water, had to give up the pursuit and to fall back on the Siberian frontier.

Between the Volga and the steppes north of Kokand, the Calmuks were, however, continually harassed by the Kirghiz hordes, who captured immense numbers of prisoners and much booty. From Kokand to the Chinese frontier, the Bouroutes or Black Kirghiz, a race pre-eminent for their fierceness and cruelty, inflicted still severer losses on the unhappy fugitives, who eventually reached their ancient pasturages numbering about 70,000 souls, or one-half of the Calmuks originally leaving the Lower Volga.

While extending her commerce with the East, the policy of Russia was therefore also directed to bringing the Kirghiz well under her domination, and, in order to effect this, she established, in 1735, a line of frontier posts along the banks of the river Ural. Putting aside the development of her Central Asian trade, there was another consideration which made Russia desirous of retaining Asiatic tribes such as the Calmuks, the Kirghiz, &c., under her control. She was possessed by the apprehension that her newly dawned civilisation ran a risk of being swept away by the invasions of Asiatic barbarians, such as those of the thirteenth century, and her policy was therefore directed to covering her frontiers with outlying populations, against whom the first waves of such possible irruptions might spend themselves. This was by no means the visionary danger it may appear to be at the present day, for not very long previous to the date in question the possibility of an advance of the Zungars had disturbed the equanimity of Peter the Great himself. Discord and civil wars brought down this great Mongol power from the height of its greatness to a state which allowed of its easy destruction by the Chinese Emperor Tsian Loung in 1756; but during many years these barbarian hordes, under their famous chief, Galdane Tsyrene, had spread terror by their expeditions against the Kirghiz of the Asiatic steppe, as far as the very frontiers of Russia.

Although the Nogai Tartars, the Bashkirs, the Calmuks, and the Kirghiz were all races admirably suited for the purposes of Russian frontier policy, it is clear that the conditions of the case demanded that these various peoples should be segregated in feelings and in sympathies, in order to weaken the influence of tribal connections, and to minimise the danger of any political coalitions being formed among them. The most watchful eye was accordingly kept on such matters, and, as has been seen in the case of the Calmuks, re-emigration eastwards was strictly forbidden to the several tribes, who were in addition kept in a state of mutual jealousy.

This policy was designed to meet such dangers

as the following: the Khan of the Volga Calmuks having, in 1701, married a relative of the Zungar chief, some forty years after, this lady sent her son with 15,000 *kibitkas*, say 75,000 souls, to rejoin her tribe in Mongolia. At that date, the Zungars had subjugated the Great Horde of the Kirghiz, and held many of the sultans of the Middle Horde as hostages; had they then made an alliance with the Little Horde by marriage, the whole of the wandering barbarians, from China to the Volga, could have been put in motion against Russia at the nod of a single chief. Such an alliance the Zungar chief actually tried to negotiate in 1750, by demanding the sister of Nourali, the Khan of the Kirghiz of the Little Horde, in marriage, but Russia naturally threw every difficulty in the way of its taking place, and it eventually fell through on account of the death of the lady in question.

From the time of the establishment of the line of military posts on the Ural, cases were frequent of this *divide et impera* system on the eastern frontier. Sometimes arms and ammunition were served out to the Calmuks to enable them to attack and punish the Kirghiz; at other times, as happened during the Pugacheff rebellion, the Kirghiz were put in motion against the Cossacks of the Ural; but the results of early Russian frontier policy are best illustrated by the rebellion of the Bashkirs.

These people, who at the present day number nearly a million of souls, first appeared in the western steppes in the tenth century, and were subjects of Russia under John the Terrible. They often rose in revolt against their Muscovite masters, and one of such outbreaks had scarcely been put down, when a religious insurrection, incited by a fanatical moollah, blazed forth with great fury in 1755. Massacres of Russians took place simultaneously in several localities, and the Tartars of Kazan joined their co-religionists and began to make plundering incursions into the heart of the empire.

With the Bashkirs to the north and the Kirghiz to the south, the feeble line of frontier outposts, stretching over hundreds of miles into Siberia, were placed in a very critical position; reinforcements of Don Cossacks and of Volga Calmuks were hurriedly sent forward, and the state of apprehension into which the Russian authorities were thrown may be judged of from what followed the suppression of the revolt, which, after all, was effected without much difficulty.

The governor of Orenburg had circulated among the Kirghiz a declaration procured from the Akhoond or Muhammedan pope of that place, which, while it declared the objects of the Bashkirs irreproachable in a religious point of view, expressed at the same time a fear lest, after rendering themselves independent of Russia, they might not subjugate the Kirghiz. He followed up this Machiavellian stroke of policy by informing these people that the Empress Elizabeth made over to them in advance, for the delivery of the fugitive Bashkirs, all the women and children of those rebels who had taken refuge over the border after the suppression of the revolt. The sensual Kirghiz hastened to take possession of the prey so cruelly abandoned to them, and the resistance of the smaller number of Bashkirs who attempted to defend their families was of course unavailing. A part of them perished then and there, others were given up to the Russians, but the majority voluntarily returned to their homes, and awaited with impatience the hour of vengeance.

Assembling afterwards in armed bands, they asked but were refused permission to pass the frontier, though secret orders were sent to the military posts, not to interfere with such Bashkirs as had commenced to cross the Ural river without the necessary permit. These bursting all at once upon the Kirghiz pasturages, commenced the avenging work of pillage and of massacre, and were allowed to continue it until the Russian governor thought his ends had been sufficiently served. Bloodshed then ceased, but the desired object of fomenting an eternal hatred between Bashkirs and Kirghiz had been thoroughly effected.

Such was the policy by which Russia eventually dominated the Asiatic populations on her eastern frontier; a policy as base as it was barbarous. Nations perhaps are subjected to the same wholesome reminders as individuals, in the story of whose lives are found actions which, as has been said, stain backwards and forwards, through all the pages of the record. And if by such bitter personal experiences, individuals are refined in the crucible and fitted to repair the errors of a misused youth, it may be that the analogous experiences of nations are needed to attain that development which may fit them to bear on the torch of progress, at the head of advancing humanity.

During the latter half of the eighteenth century,

very slow progress was made in subjugating the Kirghiz, the frontier was more or less in a chronic state of warfare, raids were made on the military posts, caravans were pillaged, and the Russians made reprisals and administered punishment from time to time. But, in the end, things became more peaceable, and Russian influence became eventually paramount with the Kirghiz; though it was not till thirty years ago that affairs were ripe for the advance across the Kirghiz steppe and the inauguration of that extension of Russian power in Asia which is watched with interest by the whole civilised world, and has a special bearing upon the dominion of England in India.

# CHAPTER VI.

## THE RUSSIANS ON THE SYRDARYA.

Comparative fertility of the banks of the Syr—Cultivation and aspects of country—Buildings on the steppe—Failures of colonists—Cazalinsk—Fort Peroffsky—Revenue of Turkestan—Population of Central Asia—The Kirghiz of the Syrdarya—Comfortable circumstances and general improvement of their condition—Their relations with the Russians — Influence of races of Turkestan on Russian culture.

THE desolation and sterility of the deserts to the north of Lake Aral grow gradually less as the post road into Turkestan turns south at about forty miles from the Syrdarya, and enters the country where commence the scattered *aouls* of the Kirghiz. These camps are pitched on patches of pasturage, covered with flocks and herds, which grow larger and more numerous as the river is approached.

It is not astonishing if, after traversing the dreary wastes skirting Aral and the sands of the Karakoom, and on reaching the banks of the river, the Russians should have imagined they had found something approaching to an earthly paradise. The country does not indeed present a very fertile aspect, but with the aid of irrigation, cereal crops give moderately good returns, fruit-trees and the vine are productive, and madder and other dye roots thrive. The mulberry flourishes, and sericulture, which is a lucrative industry on the upper courses of the river, promises to be a success in the government of the Syrdarya, judging from the experiments which have already been made. The frequent nomad encampments and the numerous cattle on the pasturages lend a charm and a life to a scene which, dismal as it perhaps really is in comparison with more favoured regions, must have seemed to the Russians, after their arduous advance from Orsk, a foretaste of the rewards awaiting them in the longed-for oases of Central Asia.

The country traversed by the Syr, in the lower four hundred miles of its course to Lake Aral, is disposed in long low undulating surfaces of clay, interspersed with stretches of sand, and dotted with occasional sand-hills. Here and there are also moister tracts of darker soil, which are covered with a brilliantly-white saline efflorescence, and, as the stream is descended, the soil seems richer and less arenaceous. It is however in spring only that the herbage can be said to be abundant, for later the summer heats dry up all verdure, and the steppe, which is generally devoid even of bushes, becomes bare, except in the immediate neighbourhood of the river. Here there is more grass, and usually a tolerably thick growth of jungle, from ten to fifteen feet in height, composed chiefly of jidda (*Eleagnus hortensis*), and of a thorny acacia bush (*halimodendron*), which is filled with pheasants. The rushes, which have already been mentioned as the special vegetable feature of the Aralo-Caspian region, occur in the pasturages upon the river's banks, and grow luxuriantly upon extensive tracks situated on lower levels subject to inundation.

The bareness of the Syrdarya country is relieved in addition to the post-houses, by a few *pisé* mosques and *mausolea* which are scattered along the line followed by the post-route on the northern bank of the river. In the Syr are many islands occupied by cemeteries full of tombs, where the wandering Turanian tribes have for ages deposited their dead. The generality of the graves are mere earthen mounds, but in some instances stone and even marble tombs are met with.

At some few points upon the route a collection of abandoned and decaying wooden shanties may be seen standing isolated, within lines of bank and ditch, from the surrounding solitude of the steppe, and are the remains of unsuccessful attempts at colonisation which have been made by Russian peasants. The nomadic spirit is strong in the nation, and emigrants not unfrequently set off suddenly, and for no apparent reason, from villages in the interior of the empire to seek a new home thousands of miles away in Asia. Such phenomena are perhaps but the shadows of those events coming, when the Russian territory in Europe shall be too small for the descendants of the present seventy and odd millions of population, which are doubling themselves every fifty years. I met several of such parties of emigrants on their long tramp to the fertile territory of Kouldja,[1] which, though being only held in trust for China, is not likely, according to general belief, to be ever given up to its real owners.

The two chief Russian settlements on the lower

[1] Here rice costs less than one penny per pound; and flour is one-fourth only of that price!

Syr, are Fort No. 1, or Cazalinsk, which is situated at about one hundred miles from the mouth of the river in Lake Aral, and Fort Peroffsky (or Ak Musjeed), about two hundred and fifty miles farther up the stream. Cazalinsk took the place of Fort Aralsk, which was founded in 1847, among the marshy tracts nearer to the mouth of the Syr, where its deserted ruins are still to be seen. Fort No. 1 is linked to Peroffsky by Fort No. 2, which stands at about three-fifths of the distance intervening between the two places. The Russians made a reconnaissance in the year 1852 of Ak Musjeed, which then belonged to Kokand, for the fierce struggle they had for its possession did not occur till the following year. The sole remains of the former fortress, which was defended by the present ruler of Kashgar, is an old ruined tower still standing within the encircling ditch and parapet of Peroffsky.

Cazalinsk is the head-quarters of one of the districts which make up the government of the Syrdarya, and this, with the government of Semiretschenk, of which Fort Vernoë is the head-quarters, formerly made up Russian Turkestan, the official residence of whose governor-general is at Tashkend. More recently, the districts of Samarcand, in the valley of the Zarafshan, and the territories of Kouldja and the Amúdarya, and quite lately all Kokand, north of the Syr, have been occupied by Russia.

Figures are not attainable which give very accurate information regarding the financial position of the Russian possessions in Central Asia. Such as exist, recall a story of a clock, which on the occasion of an august personage's visit to a certain Russian town, was at the last moment remembered to be out of order. The difficulty was got over by putting a man inside the works to turn the hands during the time the august personage remained in the town. As I have never met a Russian who believed in the statistical tables published by the Government, it is probable that the information they afford is usually something like the time of day which was shown by the town clock mentioned. The official accounts[1] have shown a total deficit of nineteen millions of roubles (say three millions sterling nearly) for the four years previous to 1872, since which date the Khivan campaign has taken place, and a further annual deficit has occurred from the occupation of the Amúdarya district. Prob-

| | Receipts | Expenditure |
|---|---|---|
| [1] For 1868 | £106,549 | £723,590 |
| „ 1872 | 323,565 | 1,204,580 |

ably the revenues may suffice to meet the more strictly civil charges of the Turkestan government; leaving the military ones to be a charge to the general revenues of Russia. Supposing the yearly cost of a Russian soldier in Asia to be 43*l.* including officers, the expense of the army in Turkestan would amount to nearly one and one-third millions sterling per annum. An entirely independent authority stated the yearly deficit to be seven millions of roubles, or say one million sterling, during the past few years, and the two amounts are sufficiently approximate to show the annual charge which the occupation of Turkestan entails upon Russian revenues at present. It is very natural that so serious a yearly deficit should act as a stimulant for the devising of means which shall restore and develop the resources of the country on which many hopes of pecuniary advantage have since so long a time been grounded by Russia.

Among notions still current, though perhaps less so than formerly, and which tend to give an erroneous idea both of the strength and of the weakness of Russia in Turkestan, is the exaggeration often met with regarding the numbers of the subject and independent populations of Central Asia. The *pres-*

*tige* of the swarming millions of the ancient Turanian hordes still clings to the locality, and in a tolerably recent work of a respectable Oriental authority, the population of Russian Turkestan, which is actually two millions—of whom a moiety are nomadic Kirghiz—has been stated at seven millions! The peoples of the three Khanates are of course inaccurately known, but their numbers are supposed to be —Bokhara, one million; Kokand, nine hundred thousand; Khiva (without the independent Turkoman tribes), three hundred thousand; so that all Central Asia, excluding Eastern Turkestan or Kashgar does not contain more than four and a quarter millions of souls! The statement that the population of British India amounts to two hundred and twenty millions is received by Russians in Central Asia with a half incredulous and a half envious air of astonishment; as well as the still more striking contrast that the whole number of British troops controlling these millions is only double that of the Russian forces in Turkestan.

The nomad population of the banks of the Syrdarya is comparatively large, though its numbers cannot be stated with great accuracy. Samorza, one of several Kirghiz chiefs near Cazalinsk, told me his people lived in 2,500 *kibitkas*, which would make this single tribe amount to 12,500 souls—a number that I elsewhere heard was an exaggerated one. However, of the one million of Kirghiz who nomadise on the great Asiatic steppe, probably more than one-third are found on the lower courses of the Syr; and after the Kirghiz and Calmuks of the Lower Volga these people are certainly richer than the tribes elsewhere. The numbers of the cattle belonging to them are estimated at—camels, 100,000; horses, 200,000; and sheep, 2,000,000. They drive a brisk trade with Russians who come yearly to Peroffsky to buy the skins of their sheep and bullocks; and they find profitable employment for their camels in the transport of merchandise between Europe, Central Asia, Siberia, and China. Their financial position being thus easy, their *kibitkas*, furs, and clothing generally are of a better stamp than usually met with among other nomads; the silks of Kokand and Bokhara, the gay chintzes and woollen cloths of Europe are worn both by men and women, while gold coins are in great demand as ornaments. These Kirghiz are well fed and good-looking, and the primitive ruggedness of their manners is evidently yielding to the softening influences of comparative luxury and wealth. Some few individuals among them have been advanced to the rank of local civil or military officers, and wear uniforms and reside in houses at Cazalinsk and Peroffsky. The carpeted rooms and the quantity of brass-bound boxes to be seen in their residences, attest the comfortable circumstances of such families, whose relations frequently come in from an *aoul* in the country, and are entertained with tea for the females, and bottled English stout and *vodka*, to which they are too much attached, for the males. The nomadic habits of the masses of the Kirghiz must, if slowly, yet surely, be in process of modification by the material advantages they derive from the domination of the Russians, and the change is making itself evident by the increasing area of land which year by year is being brought under cultivation upon the banks of the river Syr. On such lands a small assessment is levied, but the chief item of the Government revenue is a charge of three and a half roubles, say ten shillings, annually upon every inhabited *kibitka* containing about five souls. In 1873 the Kirghiz of the Lower Syr made a voluntary contribution of one hundred thousand roubles in cattle, both of carriage and commissariat, for the use of the troops during the Khivan campaign, and the circumstance gave rise to some hostile criticism and controversy. But without entering upon this or without attaching any very exceptional importance to the not unnatural assistance so rendered by the Kirghiz towards the discomfiture of their former Uzbek oppressors, it is beyond question that at the present day the relations subsisting between the Russians and their subjects on the Syrdarya are on the most friendly footing.

The terms of familiarity which exist between the Russians and the tribes of West Turkestan are somewhat striking to an Anglo-Indian observer, and however creditable such a state of things may be to the ruling race, it seems to imply that the Russian *prestige* in Central Asia is less than it might be. Though the manners of Turanian peoples are certainly rough, if not indeed ill-bred, as compared with those of Aryan races, there would appear to be something more than mere roughness observable in their demeanour. For example, I noticed a sailor on board the 'Peroffsky' steamer, on the Amúdarya, 'fend off' a native's boat from the paddle-wheel, upon which the Uzbek boatman immediately threatened the Russian (who was doing no more than his duty) with the pole he was using as an oar. This circumstance, it is to be noted, took place during the first year of the occupation of the Amúdarya district, after the campaign of Khiva; and perhaps its explanation may be found in the personal pride of ancestral descent, which is one of the chief characteristics of these Mongol-Turk tribes, when brought into relations with a race among whom this distinctive trait is wanting.

As for the Kirghiz, I have seen one of these people sitting in the *kibitka* of a Russian district chief, eating melons, and throwing their rind and seeds, unnoticed, all about the tent, though its open doorway close beside him invited the discharge of this residue into the open air. There has been some misplaced touchiness displayed in Russia (chiefly by the apostles of what is called the liberation of the Sclave populations of the East of Europe), regarding the existence of the Turko-Muscovite race. This is supposed to be a myth expressly invented by the detractors of the subjects of the Czar; but it is beginning to be acknowledged pretty generally that it is impossible to understand the country without admitting the great influence which Tartardom has had upon it in the past. Whether the development of the empire in Asia is not likely to exert a similar and very marked influence upon that section of Russians who are withdrawn from the reach of European culture, is a question which is strongly suggested in observing their intercourse with the races of Turkestan, and in noting the comparatively small interval which divides them, as far as the small delicacies and conventionalities of civilisation are concerned. Russians meet the Kirghiz more than half-way in these matters, and are acted upon rather than the opposite. That the punctilious and polished Mongol of Hindustan should have so recently come from the same blood as that of the uncouth barbarians of the Central Asian steppes is a wonderful tribute to the assimilating powers which have been exercised by the fastidious and mild Hindoos.

# CHAPTER VII.

## THE RUSSIANS ON THE SYRDARYA—*continued.*

Early summer on the banks of the Syr—Cazalinsk, or Fort No. 1—Its streets, houses, and bazaar—A Russian battalion—Characteristics of Turkestan troops—Religious and patriotic feeling—Dress—Numbers—Infantry—Artillery—Cossack cavalry—Sailors of Aral flotilla—Military alertness in Turkestan.

THE traveller who has left the Volga in early spring, when half-melted snow still lies on the ground, will, some ten days later, find a different state of things to exist on the Syr, as he enters Cazalinsk, in a thick cloud of dust, kicked up from the dry steppe by the galloping team of his post *tarantasse.* The summer heat will have already set in, and the perspiration which furrows the accumulated filth of the journey will make him look forward with redoubled thankfulness to the neighbouring shelter, where he may achieve his long-neglected ablutions in the Russian vapour-bath, whose *raison d'être* he is beginning to comprehend. He will not, however, if he is wise, and has taken warning from his experiences at Orenburg, expect too much from the little *gostonnitza* of Cazalinsk, where the bites of the venomous steppe-mosquito from which he is suffering will be fully equalled in painfulness by those of animals of a similar bloodthirsty nature which make many Russian hotels their favourite head-quarters.

Cazalinsk, whose size is about double that of Fort Peroffsky, may one day be an important place, but meanwhile this little Central Asian town of 1,500 people, does not present a very imposing appearance. Approaching it from the north a few windmills are scattered on the open steppe to the right hand, and one enters a cluster of long low houses built of raw bricks and flat-roofed, which covers perhaps a square mile of ground, and is intersected by streets running north and south, crossed by others at right angles. The houses generally have very little vegetation around them, though a few young poplar trees afford some shade, and offer by their greenness a grateful variation to the monotonous brown colour of the surrounding walls and buildings. There is also a somewhat neglected public garden, rather thickly planted, which occupies the centre of the square, in front of the hotel; and this contains a small domed shrine, which was erected by public subscription, as a memorial of the escape of the Emperor from assassination in 1866.

The doors and windows of the Cazalinsk houses are of rather primitive construction and of limited dimensions, from considerations of finance, since all wood and iron work has to be brought from Orenburg at a cost of about 15*l.* a ton for carriage.[1] The house of the chief of the district as well as the church, are of a superior description, and are built of a compact and well-burnt buff-coloured brick. On the latter building a fine dome in carpentry was being erected at the time of my visit, but the usual style of roof east of Orsk is a nearly flat one, and is formed of young poplar poles covered with a layer of rushes and a thick coating of mud over all. The whole mass makes a cool and very serviceable roof in a climate which is warm in summer, and where the rainfall is small even in the autumn, when it reaches a maximum.

The streets running north and south debouch at the latter extremity on an open esplanade upon the right bank of the Syrdarya, where Fort No. 1 is placed. This is an earthwork, surrounded by a substantial parapet and a broad shallow ditch, which can

[1] It is estimated that a round shot fired in Central Asia costs Russia twelve roubles, say nearly 2*l.*!

be filled during the floods of the river. On the side of the Syr is a bastioned front one third of a mile in length, upon which the fort is traced in the shape of a half star, which stretches in alternating long and short rays, in the direction of the town. From the buildings it is separated by an open grassy space of about two hundred yards, while on the east and west faces, the ground is open and occupied by cultivated fields. Though the ordinary garrison of Cazalinsk is no more than five hundred men, the barracks within the fort could perhaps accommodate four times that number of souls, and with the armament of field-guns which are mounted at its salients, Fort No. 1 could never have been in much danger of capture by the opponents of Russia in Central Asia. There are wells of good water in the interior of the work, though the close neighbourhood of the Syr dispenses with their use; but the water in those of the town is said to be unwholesome and brackish. Fort Peroffsky, if smaller, is a work precisely similar to Fort No. 1, though its interior looks more neatly kept, and has besides a pretty avenue of trees. It is the centre of agricultural dealings on the Lower Syr, while Cazalinsk, on the other hand, is the chief *entrepôt* for the Central Asian trade and the meet-ing-place for the agents of Russian commercial houses with the traders of Kokand, Bokhara, and Khiva. Cazalinsk also presents some animation from the numbers of Kirghiz who frequent its streets and bazaar, and who reap considerable benefit from the fact of the troops in Turkestan being allowed more liberal meat rations than the rest of the Russian army; and besides the shepherds, who drive their sheep into market, parties of mounted Kirghiz are met, for whom the dram-shops have probably an irresistible attraction. In the streets may also be frequently seen a ricketty old *tarantasse* which has come lumbering in from the adjacent steppe, and carries some Kirghiz ladies to town to do a little shopping. In their quaint white upright turbans, and gaudy chintz or silk gowns, they offer a picturesque sight and are often good-looking. For Muhammedans, they enjoy a considerable amount of freedom, and, according to general report, very seldom abuse this privilege.

Though there are a few shops in different parts of Cazalinsk, the bazaar square, whose general aspect is decidedly Oriental, is the chief centre of traffic. Here many of the *boutiques* are kept by Russians, who sell cigarettes and goods of European manufacture, of rather inferior quality; but the majority of the traders are either Tajiks or Central Asian Jews, who deal in gaudy chintzes, in the Kirghiz taste, or in curiously patterned silks of Kokand or Bokhara. On one side will be seen piles of bar iron from Russia, on another the willow-wood frameworks of the *kibitka* of the nomads, while on a third side is a row of butchers' shops. Raw cotton, dried fruits, and hides are the chief staples of trade, and strings of camels employed in the transport of these goods are generally stationed in the bazaar square, where are also picketed the horses of the Kirghiz who have come into the town for the day.

On my arrival at Cazalinsk, the esplanade round the fort was covered with lines of little *tentes d'abri* of a regiment of Turkestan rifles, *en route* to relieve a portion of the force in the Amúdarya territory. The battalion in camp was a fair specimen of the Russian troops in Central Asia, and was composed of first-rate material; the age of the men seemed to be between twenty-five and thirty-five years, their appearance was excellent, and in height and *physique* they were all that could be desired, though their spare and robust figures might not perhaps have been quite 'set up' enough for a British eye. Their conduct, as far as a curious stranger like myself could judge, was good, and their discipline above praise; and if I was told they were not entirely inaccessible to the charms of the *vodka* bottle, I am bound to add that, during six months in which I had a constant and close opportunity of observing them, I never saw a soldier in Turkestan the worse for liquor. Their mornings and afternoons were pretty well taken up with rifle practice and with drill, and their evenings were devoted to dancing and chorus singing, both of which recreations they enter into heart and soul, and enjoy thoroughly.

Among the ranks of the Russian army that sentiment is carefully fostered which Montesquieu has pointed out as characterising the ancient Roman legions; and with the love for their country and respect for their rulers with which the Russians are impregnated from their youth up, a deep religious feeling is always found mingled. Before their meals the men fall in to chant a short grace, and at tattoo the musical rhythm of the prayer for the Emperor and the evening choral hymn has an indescribably fine effect, as it swells through the surrounding silence and solitude of the deserts of Turkestan.

The system of mental training and preliminary

education by which soldiers are laboriously prepared for the day of fight, and which, in England, has taken a form not inaptly termed 'buckle-polishing,' does not hold a very high place in Russian military estimation; though the 'mujik,' who has hitherto supplied the ranks of the Russian army, is generally supposed to be a very inferior being, intellectually, to the British private. It is certainly true that a course of buckle-polishing would add immensely to the smart appearance of Russian soldiers, for there are few of them in Turkestan who would not provoke the choler of a British sergeant-major, and even the Imperial guard would scarcely do more than pass the inspection of the same military Rhadamanthus. Yet, in the absence of this conventional education, there exists in the breast of the Russian soldier some strong moral principle, which gives him a discipline and a constancy to support hardships and fatigues with a smile, in repeating the simple formula, 'Service of the Czar.' Whether this devotion is merely due to a reprehensible ignorance of the demands of the scantily-supplied Russian labour market, or to the possession of an inferior intellect, which is incapable of making 'any complaints,' it would be needless to enquire, since its result is sufficient to extort the admiration of the most prejudiced observer; and though a point cannot be made in favour of the Russian soldier by saying he does not desert (for desertion would be all but impossible to him), it is certain that he fights without flinching and dies with resignation, upon uncommonly small pay and very scanty rations of coarse food.

His dress in Asia is a very practical one, and consists, in summer, of a white cotton blouse, approaching to a Norfolk jacket in form, and ornamented on the shoulder-straps with the regimental number in coloured figures of embroidery. The cotton blouse and the soft cerise-coloured leather trousers he wears are both of local manufacture, from financial considerations, and the latter articles of clothing are comfortable, cheap, and serviceable, and are worn tucked inside high boots both by officers and men. In the winter, the green cloth tunic replaces the white cotton blouse; and the usual grey overcoat, buttoned in at the waist, is also worn, with its accompanying *bashalik*, or hood; which is an indispensable muffler for the head and ears, while the cutting freezing winds of the Turkestan steppe are blowing. In the summer, the *képi*. with its white cotton cover, seems an insufficient protection for the head, but sunstrokes are nevertheless rare, and seldom happen except under special circumstances, as in such forced marches across deep sandy deserts as took place during the advance on Khiva in 1873.

The Russian forces in Turkestan number nearly twenty-six thousand men, of whom the infantry comprise about eighteen thousand, and include four battalions of rifles, armed with the Berdan breech-loader, an arm which is not yet supplied to the remainder of the troops. In the number are also several garrison companies, who occupy forts and military posts, and who are more or less relieved from the duties of active field-service. The artillery numbers about two thousand men, attached to two mitrailleur batteries and forty Cossack field-guns, of four and nine pounders, as well as to forty heavier pieces of position, chiefly in the forts. At Fort Vernoë, the head-quarters of the government of Semiretschenk, there is a Cossack colony, which is supposed to supply twelve squadrons, and there is perhaps an equal number of cavalry in the Syrdarya government, which are furnished from the Orenburg and Ural Cossacks. The popularity of this service in Central Asia is not so great as it used to be in the old days, when frequent frontier raids, that lasted but a limited time, were made on the nomad *aouls* of the steppe, and much plunder was brought back to the villages in the valley of the Ural. At the present day the squadrons spend some years away from their homes, and are literally in camp the whole time they are absent: a life which entails losses in horseflesh they replace for themselves, and during which no plunder is now obtainable. These horsemen are armed with a cut-and-thrust sword, long rifle and bayonet, and merit rather the description of mounted infantry than of irregular cavalry. They are mounted on stout roadsters, standing about 14½ hands high, and shaped rather for carrying weight over long distances than for speed. These horses are bred in the Orenburg and Ural countries.

The military forces of Russia in Turkestan include a battalion of about five hundred and fifty sailors, who man the vessels of the Aral flotilla, and, though most of them are villagers of the Ural valley and entirely unacquainted with the smell of any salt water save that of the brackish expanse of the great Central Asian lake, such a shortcoming does not prevent them from being equally as fine fighting material as the soldiers. The high priests of the mysteries of holystoning would, perhaps, laugh at decks being washed by two men in long cavalry boots, one using a mop, the other dribbling out water from a small tea-kettle; but the capacity of endurance and joviality under hardship displayed by the sailors of the Aral flotilla, at least command the respect of more nautical peoples.

What perhaps most strikes an observer who is accustomed to the charming military *insouciance* of an Indian cantonment, with its mock paroles and countersigns, is the state of activity and readiness which distinguishes the forts and camps of Russia in Turkestan. A field force could always start at a few hours' notice from one of the positions which form the bases of military operations; while in the standing camps, such as those of the Amúdarya, matters are conducted strictly as in an actual campaign, and advanced pickets are nightly thrown out to bivouac on all positions by which an enemy could approach, although there may be none within hundreds of miles of the place.

# CHAPTER VIII.

## THE SYRDARYA.

Attenuation of Syr in its lower course—Flow of Syr or Jaxartes—The plain on the east of Aral—Probable ancient course of Syr—Present terminations of the rivers Chuy, Sary-Su, and Talass — The changes of outlets of the Syr—The Jánidarya and other historical arms of the river—The Asiatic Tanais of Alexander—Facility for changing course of Syr—Possible causes of the physical aspects of the Kizzelkoom deserts — Influences of human action upon surface of the earth.

THE Syr, which is a river fourteen times larger than the Thames is at Staines,[1] is far from giving this idea on inspection, and the fact can only be fully realised after considering the large portion of its volume which is utilised in the cultivation of the country it traverses, as well as that which is actually wasted in marshes along its lower courses.

Flowing in a general direction from east to west, the Syrdarya or ancient Jaxartes takes its rise in the

[1] Flood discharge of Thames at Staines, 400,000 cubic feet per minute ; probable flood discharge of Syr, 5,500,oco cubic feet per minute, of which quantity less than one half enters Lake Aral.

high valleys of the Russian territory in Central Asia which lies to the south of Lake Issikkul. From its sources, it passes four hundred miles to the west, when it enters the Khanate of Kokand and crosses that country in a south-westerly direction, along a distance of about three hundred miles. Entering Russian territory again, it soon makes a remarkable turn, and flows nearly due north, four hundred miles, after which, changing its direction more and more, it meanders north-west, along a distance of about four hundred and fifty miles, to its outlets on the north-east shore of Lake Aral. Its total length may thus perhaps be between fourteen and fifteen hundred miles.

Though it is common enough to speak of the valley of the Syrdarya, the expression is scarcely an accurate one, for when the river emerges from the hilly country near Khodjend, it enters upon the great plain sloping down from the east towards the basin of Lake Aral, and possessing no definite watercourse line. The question whether the Syrdarya formerly flowed directly to the west, and has changed its course just below Khodjend at the remarkable turn which has been mentioned, is a highly important one. A glance at the map will show that the shorter and more natural line of the river to an outlet would have been almost due west across the country now occupied by the Kizzelkoom[1] desert, and any change of the kind alluded to would have deprived the affluents reaching the left bank of the ancient Syr of an outlet for their waters, which would thus have been lost in the sands, instead of entering Lake Aral. It will also be seen that the rivers Chuy, Sary-Su and Talass, which now end in sands and marshes to the north of the Syr, should have naturally had an outlet in Lake Aral; where other rivers which formerly watered the Kizzelkoom would also have disembogued. Thus the lake has been deprived of much of the water which anciently flowed into it, and the diminution of its waterspread and the desiccation of the surrounding country will be partly accounted for. Plains have been covered, while valleys and watercourse lines have been choked by the sand and gravel formerly carried down to a receiving basin; habitable ground, which was clothed with herbage and cultivation, has been dried up, vegetation and forest have died off; climate has been modified as moisture has disappeared, and

[1] Kizzelkoom, i.e. Red sands. These deserts have an area of about 35,000 square miles.

barrenness and desolation have replaced the smiling and bounteous aspects of Nature in regions which history tells us supported enormous populations.

The causes which have changed the direction in which these rivers formerly ran from the east towards Lake Aral, as well as those which have arrested their discharge altogether, will be discussed in future pages; but, meanwhile, a few facts may be stated which afford strong reason for thinking that these rivers generally have been subjected to the same phenomena as those which history tells the stream of the Oxus or Amúdarya has been characterised by since the remotest antiquity.

In recent times the Syrdarya in its lower courses has been subjected to frequent change. From the neighbourhood of Fort Peroffsky a dry bed of the river called the Jánidarya leaves the left bank, and runs south-west across the Kizzelkoom desert into the dry lake of Kouktchatengis, from which again other dry channels run towards the Lower Amúdarya and towards the south-east corner of Lake Aral.

A few miles below Fort Peroffsky the Syrdarya now divides into two branches which reunite at about one hundred miles to the west, near Fort No. 2, and the country included between these two arms of the river comprises a vast tract of swamps, where much water is lost by evaporation. The division is said to have been caused, some fifty years ago, by the opening of a canal of irrigation from the right bank; and the rush of water into this canal which is called Karauzak, and whose head is nearly opposite to that of the Jánidarya, enabled the flow by the latter arm of the river to be easily stopped. The diversion of the stream into Karauzak has seriously deteriorated the other branch of the Syr, which has, in consequence, received the name of the Jamándarya, or bad river, from the number of shoals that have been formed in its channel.

From the Jamándarya, another arm of the river is given off to the south, but this soon changes its initial direction, and flows westwards into Lake Aral, parallel to the lower course of the Syr. At the present time no water passes along this course, which is called the Kuwándarya, except perhaps a very limited quantity during extraordinary floods. Such water, however, does not now reach an outlet in the lake, but moistens large tracts of country covered with thick growths of rushes.

By the Jánidarya course Mouravieff saw the waters of the river flowing in 1816, soon after which date, it is said, the Kirghiz barred its head and so prevented an entry of water into it; but in 1740 Mouravin, who was in these countries, saw no water passing along this arm. Maps of more than a century ago show the Syr and Kuwán as separate rivers, and Russian information of the end of the sixteenth century tells us that the mouth of the Syr was in the Blue Sea, *i.e.* Lake Aral, and that this river received the Kenderlik, a stream not identified, and which is supposed to have disappeared. From the same source we also learn that the Sary-Su stopped short at that time in marshes before reaching the Syr.

The Emperor Baber, who conquered Hindustan, says that the Syr, after flowing to the north, lost itself at the commencement of the same century in the sandy desert; and perhaps from this it may be concluded that its mouth was then situated somewhere north of the Karakoom sands, and that its waters may have added to the extent of the basin which now receives the streams of the Irgeez and Tourgai. At any rate, the map in Ortelius, 'Theatrum Orbis Terrarum,' that illustrates Jenkinson's travels of the sixteenth century, shows the Syr flowing into the Kitay Lake, which is placed at some distance to the north of the termination of the Oxus, and from which the river Obi is also shown as taking its rise.

In the fifteenth century it is tolerably certain that the waters of the Syr had an outlet in Lake Kouktchatengis, on the south-east of the present waterspread of Lake Aral, and flowed on from there by the Kizzeldarya course, towards Kunya Urgenj, upon the old course of the Oxus; though in the fourteenth century, according to Abulfeda, the Syr had one outlet at the south-east and the other at the north-east of Lake Aral. That the river, or at least some portion of it, passed in a south-westerly direction in the fourteenth century, is almost certain, since the caravan route from Urgenj to Otrar, near the junction of the Arys with the Syr, followed this line. At that date, however, the point of bifurcation was probably situated more to the east than where it is now found. Shumsuddín Dimashkí records the tradition in his writings of this date, that in Sogd twelve thousand canals were derived from the Sihún or Syr, a number which was equal to that of Alexander's generals. However exaggerated the statement may be, it may stand as an indication of the former flourishing state of irrigation in the countries east of Aral. A passage in Abulghází

Khan's memoirs seems to imply that in the thirteenth century the Chuy and Sary-Su fell into the Syr; though the Talass was then lost in the sand, according to the information given by Rubruquis. In the twelfth century Idrísí tells us the mouth of the Syr was only thirteen miles from that of the Oxus, which would put it at the south-east of Aral; into which also two other rivers emptied from the east. In the tenth century Masúdí, among other rivers reaching Lake Aral from the east, names, in addition to the Syr or river of Fergхána, a great stream called Barak (Terek, according to Ibn Haukal); and on this boats carried different kinds of merchandise to the lake. Balkhi-Istakhrí states, in the same century, that the outlets of the two rivers Amú and Syr were three hundred miles apart, which would place that of the latter far to the north. Six centuries previously Ammian Marcelline speaks of the two rivers, Araxetes (probably Jaxartes) and Demas, as flowing from the Sogdian hills, across the plains, and carrying boats to the long and broad Oxian marsh, and these rivers thus reached Aral on its eastern shore. The latter seems to be identical with the Demus, which is mentioned, together with Biscatis, by Ptolemy in the second century as affluents to the left bank of the Jaxartes. The mouth of the old Syr appears, according to this geographer's information, to have been at the south-east corner of Aral. Between the second and fourth centuries there therefore seems to be an indication that the Jaxartes changed its flow in its upper courses, and that Biscatis was lost while Demus was left to flow on separately to an outlet.

According to Strabo, in the first century of our era the mouth of the Jaxartes was in the Caspian, *i.e.* in that united Aral and Caspian Sea which has been spoken of in a previous page; and the eighty parasangs, or three hundred and fifty miles, which he states [1] (on the authority of Patrocles) to separate the mouths of Jaxartes and Oxus, approximates very nearly to the distance measured from the south-east corner of Aral, to the old Oxus mouth in Balkhán bay.

We know, from Arrian's history of the expedition of Alexander, that Cyropolis was situated at a distance of rather more than three days' forced marching from Samarcand, and on the banks of a river whose bed was dry at the moment the former city was

[1] In another passage of Strabo, this distance is stated to be 2,400 *stadia*.

stormed. The interval that must thus have separated them affords grounds for thinking that Cyropolis was too near to Samarcand to allow of its having been on any point of the present course of the Syrdarya; and indicates its position to have been rather on another river, which flowed to an outlet in Lake Aral by a course following that which has been mentioned as the more natural one for the Syr. In fact there is, at the present day, along the south of the Kizzelkoom desert, a chain of small oases and of salt lakes, that perhaps represents the old channel of the Tanais, on which were the sites of Cyropolis and the other five cities destroyed by Alexander. Ten years ago, Glukhoffsky, when on his way to Bokhara, passed between Chináz and Djizzak, across the easterly portion of this line, and travelled through large tracts of rushes, which, according to local tradition, occupied an ancient bed of the Syr. It may be mentioned that, in October 1874, a canal was commenced to be excavated for the purpose of carrying a portion of the waters of the Syrdarya in the general direction of this old course, and it is quite possible that the opening of the canal may have a serious effect upon the present flow of the river, and even perhaps change it once more to an outlet in the south-east corner of Lake Aral. The different circumstances which have been mentioned afford strong grounds for thinking that the several rivers crossing the plain on the east of Lake Aral have had courses which were at various times connected with one another, and have changed their outlets frequently, from the phenomena that characterise their flow.

Generally speaking, it may be said that the flow, or the cessation of flow, of the waters of the Syr-darya, by its different branches, are under the control of artificial means, and that the intelligent application of a very moderate amount of labour would turn the river into any direction across the Kizzel-koom desert, short of one actually running up-hill. The operation is facilitated by the circumstance that, during more than one-half of the year, the volume of water flowing in the channel is at least from three to four times greater than that of the remainder of the year. At the epoch of minimum flow, the Syrdarya is therefore a shallow stream, with a low velocity, and a canal can then be excavated in such a position as would ensure a large body of water passing into it on the advent of the succeeding floods. The set of the river having thus been influenced, no long time would elapse before the whole volume of the stream would change its direction into the artificial course, and flow to a new outlet.

The frequent dry beds of irrigation canals, the numbers of ruined fortresses, of mausolea and of tombs, which cover the country on the banks of the Syrdarya, show that in former times the population was more numerous and more sedentary than at the present day. Such remains and antiquities as have been examined do not apparently belong to epochs more distant than those of the Arab and Mongol dominations. Some are Muhammedan, while others are Buddhistic, but relics of ancient Greek and Persian origin remain still to be disinterred from the sands of the great unexplored Kizzelkoom desert, where they are probably buried. From these it may be hoped that some day the historian and the archæologist will extract revelations of the deepest interest.

Nor are the traces of ancient peoples and of past principalities alone perhaps to be recovered from these all-devouring deserts; for should the studies of physical geographers resolve the problems which are involved in the changes in the course of the Syr, the disappearance of its historical affluents, and in the present condition of the Chuy, the Sary-Su and the Talass, a ray of light may be thrown on the obscurer and earlier days of the human race. The chaos of dried-up river beds, of salt marshes, and of far-stretching sand wastes, may be found to be but an antetype of the catastrophe which overtook European civilisation in the fifth century, and of the destruction which swept over Roman society on the descent of Attila and his Huns. The Asiatic wastes remain unreclaimed, while the Europe of the Huns has been renewed and restored, and thus the wreck and ruin of nature, the dry bones of a once fecund earth, which strew the deserts of the Kizzel-koom, may perhaps be due to a want of brain power and to an incapacity for scientific observation, that rendered the Turanian hordes powerless to appropriate and continue the civilisation which was current among the Aryan races, whom they swept away. But, on the other hand, it may be that the desolation of these deserts is a relic of the battle of the infant human intellect with the as yet uncomprehended phenomena of nature; of a struggle which has indeed left ruin behind it, but in which were nevertheless gleaned the first rudiments of that knowledge which has produced the modern art of irrigation, and which may once more restore culture and productiveness to the wastes of Turkestan.

In any case, we have here a wide field for the observation of that power which is exercised by man upon the earth, and which an eminent Italian geologist treats as a new physical force, wholly unknown to earlier geological epochs, and not unworthy to be compared, in its energy and universality, to the most appalling effects of the volcano and the earthquake upon the surface of terrestrial nature.

# CHAPTER IX.

## THE BASIN OF LAKE ARAL.

Inconvenient navigation of the Syr—Its deterioration as a navigable channel—The Aral flotilla—*Saksaoul* and coal fuel—Banks of river below Cazalinsk—Population—Cultivation—Aspects of lower courses and mouth of Syr—Kos Aral—Appearance of Aral—Description of the basin—Area—Depths—Evidence of fall in the surface of the lake—The conditions of its intermittence and actual existence—Its overflow and entire desiccation.

CAZALINSK derives much of its importance from being the head-quarters of the Aral flotilla, whose dockyard is under the guns of the river face of Fort No. 1. It is about twenty years since the first vessel for the navigation of the Central Asian waters was carried in pieces across the steppes from Orenburg, and put together upon the Syr; and, in spite of many drawbacks, there can be no doubt of the great services which this naval establishment has rendered to the progress of the Russian power in Turkestan. The difficulties of navigation in this part of the world recalled to my recollection the sight of a steamer, stranded far inland, which I saw years ago as I was passing down the Ganges, and which awaited a high flood to enable her to be once more placed upon her natural element. And, although the Ganges is between seven and eight times a larger river, the Syr is also liable to somewhat similar inconveniences, since at least on one occasion a Russian vessel has only been saved from an analogous mishap by the goodwill of the Kirghiz, who gave their labour to floating her off the sandbank on which she had grounded.

As a navigable channel, the Syrdarya has scarcely fulfilled the expectations which were formerly entertained of its forming a convenient line of communication with the heart of Turkestan; although the Russians long ago closed several irrigation canals, in order to increase the stream of the river. During the flood season, powerful engines are required for the vessels to overcome the force of the current, and though this inconvenience is diminishing on account of the larger volume of water which is yearly being diverted for cultivating purposes, it is replaced by a still greater evil, that threatens to do away entirely with any value the river still possesses for steam navigation. From the point of bifurcation of the Jamándarya and Karauzak, down to where these arms reunite, the bed of the former arm is obstructed by a mass of shoals, due to the division of the volume of the stream, and the consequent deposition of earthy matters, and it is now passed with great difficulty by the vessels of the Aral flotilla. This deterioration was sensibly increased in 1874 by the opening of a large canal of irrigation in the territory of Kokand, and the future action of the canal now being dug near Khodjend, must have a similar injurious result. The conclusion therefore is that nothing but a very elaborate and expensive engineering project, which should assign a proper share of the waters of the river for the purposes of irrigation and navigation, will ever succeed in making the Syr a convenient line of communication with Turkestan; and its absolute abandonment as a navigable line and the utilisation of its entire volume for purposes of irrigation is in consequence strongly suggested.

The Aral flotilla comprises some half-a-dozen paddle-wheel steamers, varying from 70 to 180 tons burden, and furnished with engines of from twenty to seventy horse power; and, besides these vessels, there are a few steam launches and a dozen heavy transport barges for the accommodation of troops. At Cazalinsk there is a floating iron caisson dock, in which repairs can be executed, under the supervision of an American engineer and his staff of workmen.

Owing to the Syr being icebound for at least three months in the year, and to the high gales which prevail during autumn and winter, the Russian vessels are on service between May and October only. During these months the summer floods provide in general a sufficient depth of water for the navigation of the river, though the current at the same time is increased to such a degree that more than three weeks are spent in ascending to Chináz, a distance of about seven hundred miles from Cazalinsk. Even with the increasing yearly diminution in the volume and velocity of the river, a current of from four to five miles an hour is experienced during the flood season, and the vessels scarcely possess sufficient engine power to permit of much way being made against the stream, especially when they have heavy transport barges in tow.

The fuel used consists of the gnarled roots and stems of the shrubby tree known as *saksaoul* (*haloxylon ammodendron*), which grows, though it can be scarcely said to flourish at the present day, on the steppes east of Aral. It is somewhat plentiful near Fort Peroffsky, where it costs about ten shillings a ton; at Cazalinsk it is twelve shillings, and on the Amúdarya, where it is scarce, the value rises to twenty-five and even to thirty shillings a ton. Its heating properties are equal to one half only of those of Tashkend coal, whose prime cost with good management would not be much in excess of that of *saksaoul*, and the use of the mineral fuel promised therefore to result in considerable economy. The promise, however, has remained unfulfilled, for the working of the coal mines has, for some mysterious reason, hitherto resulted in failure, though a handsome fortune probably awaits anyone with the energy, the capital, and above all the talent for business matters, which are required to develop the mineral capabilities of Turkestan.

As a mere financial enterprise the working of the Aral flotilla can scarcely be considered a success, for figures show the cost of such transport operations as it effects to be eight shillings per ton a mile; while private merchants actually pay about sevenpence per ton a mile, and the army *Intendance* perhaps fifty per cent. more than this last rate. These circumstances have suggested the abolition of the Aral flotilla, since the demands made upon it for merely military services upon the Syrdarya at the present day are very few, and since the peculiar conditions of the Amúdarya require a special naval establishment, whose foundation has already been laid.

The Syrdarya at Cazalinsk has a breadth of less than one thousand feet, with a maximum depth of twelve feet; but these dimensions decrease considerably as it is descended to its mouths in Lake Aral. In June 1874 the river appeared to be in moderate flood, though so much of its volume is lost in the Karauzak marshes, and is diverted by many small irrigation cuts, that it was only six feet deep, at the most, in the largest one of the three mouths by which it enters the lake. The navigable channel becomes more and more tortuous as it is descended, and the 'Samarcand,' in which I made the passage, frequently touched the ground at the bends, though without doing herself any damage, as the bottom was tolerably soft. The country along the lower courses of the Syr has little elevation above the stream, which is bordered on both sides by rushes, and which meanders through pasturages where the numbers of cattle of every kind struck me as being large, when it was considered that the majority of the Kirghiz had already gone north with their flocks and herds for the summer. As the steamer passed down, groups of horses and cows assembled on the banks at short distances, to look with affrighted eye and distended nostril at the smoking 'devil ship,' as these vessels are called by the population of Turkestan.

The cultivation practised hereabouts by the Kirghiz is of a primitive and limited nature, and all details of agricultural labour fall to the lot of the poorest individuals, who look after melon beds or small plots of cereals, which are watered by means of a large wooden scoop, suspended from a triangular gyn erected upon the banks of the river, and having a long handle that is worked by two men. Near the mouth of the Syr are larger areas of ground, which are ploughed by means of bullocks, and are fertilised by canals conveying streams to these low levels from the river.

From Cazalinsk for fifty miles downwards there are only a few clumps of trees, chiefly of a kind of willow. This timber ceases entirely about the low sandstone ridges lining both banks at a mile or two of distance, down to Lake Aral, whose shores they certainly formed at a previous epoch, when its waterspread had a higher level than now. Excepting an occasional mud mosque, and the deserted and ruinous barracks of Fort Aralsk, there are no buildings to be seen along the lower hundred miles of the Syr; but the frequent small ferry-boats and the many *aouls* which are dotted over the country, show that the locality is a favourite one with the Kirghiz. The actual numbers seen from the deck of the 'Samarcand' were indeed large, and the rags and the dirt of both sexes showed that they were among the poorest of the tribes, and that the richer families were absent on their summer wanderings.

The Syr at its mouth has filled up an extensive estuary with sand, and the size of this has also been added to by the gradual fall which has taken place during the past century in the level of the lake. The locality in question is a low rush-covered tract covered with large pools of water, which are crowded with aquatic birds, such as storks, pelicans, &c.; and on the drier ground stands a semaphore for signalling vessels, besides some huts, forming a kind of depôt where *saksaoul* fuel and naval stores are kept for the use of the flotilla. These huts are also tenanted by a few Russian sailors and semi-sedentary Kirghiz who act as pilots, and who catch large quantities of a coarse sturgeon, which are plentiful in these waters, and from which oil is extracted. Opposite to the estuary is the island of Kos Aral, whose sandstone cliffs acquire an appearance of exaggerated boldness from the flatness of the depressed marshes across which they are viewed.

Even for the matter-of-fact minds of our day, there is, apart from the remoteness of its situation, a veil of mystery enveloping Lake Aral, which is sufficiently provocative of the interest of the most phlegmatic observer. That the most ancient classic historians should have spoken of countries which are situated farther to the north and to the east, yet should have omitted all mention of an inland sea which is one hundred times larger than the Lake of Geneva, is an enigma stimulating enough to demand a solution. Nor is the attraction of so curious a circumstance lessened in considering the fact that, though the existence of the lake remained actually unknown to Europeans until comparatively very recent times, the Arabians who conquered Central Asia had no difficulty in observing and recording its exact dimensions more than ten centuries ago.

My first view of the Great Khwarezmian Lake showed it to possess a special aspect which harmonised, as it were, with its strange historical attributes. A low far-stretching swamp, covered with vivid verdure, and interspersed with patches of yellow sand and shining azure pools, formed the foreground, where water-fowls were sporting. This separated the river from a waveless green expanse, with limits of deeper tones which stood out upon a sky of curdled milky blue; while in the middle distance, and cutting the horizon sharply, rose the dark violet mass of Kos Aral, whose steep sides were mirrored perfectly in the clear waters beneath them. Though the sun lit up streaks of fantastic colour or glanced on snow-white plumage, and though the monotony of the surrounding deserts was for a moment forgotten, solitude remained the pervading spirit of the scene; and when night closed in, this sea without ships and these shores without ports assumed a sadness which was all their own, as a nebulous mist veiled the starlight and spread like a pall over waters whose ripples gave out no gleam of phosphorescence.

The Sea of Aral is a sheet of water having an area of about 24,500 square miles, and thus covers an extent of ground which is very little smaller than Scotland. It occupies the north-eastern part of the great Aralo-Caspian depression, which Humboldt ascribed to a rupture and consequent sinking down of the crust of the globe. Its waters,[1] though sensibly salt, are under necessity potable by the antelope and domestic cattle upon the larger islands, and the saltness is locally modified by winds driving before them the streams of fresh water discharged by the Amú and by the Syr; the sole two rivers emptying into the lake at the present day. The deepest part[2] is found in a strip along the western shore, immediately under the cliffs of the Ust-Urt plateau; and there is a second less deep strip, running north and south, down the centre of the lake. From here the water shallows, though not very regularly, to the north, south, and east shores, whose gentle slopes are occupied by rushy marshes, among which the edges of the lake are lost, and where sand dunes afford one among other indications of a fall in the level of

[1] The water of Lake Aral contains 13 only of salt in 1,000 liquid parts; that of the Atlantic contains 42.

[2] The greatest depth is 37 fathoms.

the waterspread of Aral. A large sandy tract on the north is shown by a Russian map of the last century to have been under water at that time, and many bare water-worn hills are also seen in this direction. Meyendorff states that at the commencement of the present century the water, which is now forty miles distant, washed the foot of the hills beyond the peninsula of Kuk Tornak, near the northern extremity of Lake Aral; while a minaret upon the eastern shore, which is now many hours' walk from the edge of the water, is known to have been formerly close to it by the greybeards of the Kirghiz. These people also say that they can at present ford the Syr at points where it used to be impassable. It may be added that Admiral Boutakoff, who was the first to navigate Lake Aral, found the traces of waves upon the cliffs of Ust-Urt and of the islands, at a level far above that which could be reached during storms at the present day. And, as will be presently described, the slopes of the Kashkanatao hills, on the Lower Amúdarya, have very distinct horizontal water-marks and rippled beaches, whose height is many feet above the level of Lake Aral. The configuration of the surrounding country shows that with a surface about sixty feet higher the waters of the lake would have overflowed at the different places which have been indicated on a previous page. Moreover, from the action of the northerly gales which blow in the Aral basin, it is tolerably certain that such overflow would finally have ceased at the head of the Abougir Gulf, which occupied the south-western corner of the lake. The bed of this gulf was covered to a depth of three feet by water in 1848, but is now dry and under cultivation.

These circumstances incontestably prove that Lake Aral has formerly been a much larger and deeper body of water, and a very few figures would show that about double the supply of water which it now receives would soon again raise its level to overflowing, while, on the other hand, if the present supply were cut off, about ninety years only would suffice to dry up the lake entirely. Replenished as this basin is with a very limited amount of rainfall, the existence of the body of water it contains practically depends on the single condition, that the quantity emptied into it by its tributary rivers shall equal that which is evaporated from its surface.[1] In historical times many rivers contributed

[1] Water contained in Aral = 1,233,434,000,000 cubic yards. Average receipt from Amú and Syr, 2,000 cubic yards per second, which

their streams to Lake Aral which now stop short in the sand, while others have changed their courses and have so deprived their affluents of a passage to the natural receptacle for their waters. From these reasons, the lake has degenerated from an extensive and tolerably deep body of water, possessed of a copious overflow, to a shallow pool, which has at certain historical epochs been shallower still, or has even perhaps dried up sufficiently to have caused its practical disappearance from the map. Such, in somewhat general terms, is the explanation, which will be developed in the following chapter, of the difficulties which have been experienced regarding the existence and history of Lake Aral in past times.

would provide for an evaporation of about 30 inches annually, from the surface of the lake. Probably the loss on this account is somewhat greater, though the level seems now to be maintained by the rainfall as well as by the earthy deposits made by the two rivers.

# CHAPTER X.

## THE ARALO-CASPIAN SEA.

The Asiatic Mediterranean—Its escape towards the Frozen Ocean—Reconstruction of this Sea—Opening of the Bosphorus—Consequent separation of the Black, Caspian, and Aral basins—Junction between the Aral and Caspian by overflow of the former—Consequent aspects of Aralo-Caspian Sea—Existence of this depended upon surface of water in Aral basin—Variations of rivers feeding Aral caused intermittence of overflow.

On a previous page a reference was made to a past epoch in the world's history when the waterspread of a great fresh-water inland sea covered the plains of the Lower Danube and of Southern Russia, and, extending eastwards and north-eastwards, included the basins of the Black, the Caspian, and the Aral Seas, and their neighbouring low-lying steppes. This idea, which was due to Tournefort, found a support from Buffon, and more than half a century ago was developed in a memoir submitted by Dureau de la Malle to the French Academy. In this memoir the separation of the three basins was explained by the rupture of the straits of the Bosphorus, which there are grounds for attributing to a volcanic commotion at the mouth of this opening in the Black Sea, and which caused the cataclysm, about 1529 B.C., known as the deluge of Deucalion. The subsequent researches of naturalists have placed beyond doubt that the *fauna* of the three basins are very nearly related, if they are not indeed entirely identical; and it will now be shown that, if the present outlet at the Bosphorus were once more closed to a height of (about) two hundred and twenty feet above sea-level, a restoration could be made of the state of things which is described in Dureau de la Malle's memoir as having actually existed. In addition, grounds will be adduced for believing that such an Asiatic Mediterranean would also have passed its surplus waters over the low transverse ridge which is on the north of the basin of Aral, and would have flooded the country about the sources of the high Tobol and the Ubagan, which flow into the river Obi. Observations have not yet been made to determine the height of this watershed, but it is supposed to possess generally more than the two hundred and twenty feet which the lowest point in it should at the most have, in order to allow of the overflow indicated. In his 'Asie Centrale,' Humboldt has, however, spoken in very decided terms of the low elevation possessed by this ridge, and there is no reason whatever for assuming the impossibility of an overflow taking place over some point in it, at the height mentioned. Still more since, as it will be seen, the different parts of the remainder of the Aralo-Caspian depression have undergone no such elevation, in comparatively recent times, as has hitherto been supposed, and since their actual heights still admit of the restoration of pre-historical conditions and the effacement of such local physical changes as have occurred in more recent times, it all but certainly follows that the ground on the north of the Aral basin has also undergone no displacement, and that, at the height of two hundred and twenty feet above the sea, the waters would pass over into the valley of the Obi in the same way as there is the very strongest presumption of their having done in early historical times, and even in the seventeenth century, as will hereafter appear.

Supposing, then, the outlet of the Bosphorus to be closed to the height of two hundred and twenty feet above sea-level, the superfluous waters of the Black Sea basin, which now flow off to the Mediterranean, would rise in level and encroach on the south Russian steppes and the lower Danube plains, though the coasts of Asia Minor, which form the southern boundary, would be but little changed on account of their steepness. On attaining a height of about twenty-three feet above sea-level[1] the waters would escape by the line of the Manytsch into the basin of the Caspian, and, after having filled it up also, would flood the country intervening between it and Lake Aral. In their ascent to this basin the waters would chiefly pass by the Emba steppes from the north-east of the Caspian basin, and from Balkhân Bay on the south-east, up the country crossed by the Uzboy channel of the old Oxus; for between the two seas lies the elevated plateau of Ust-Urt. This high ground has several detached portions near the Caspian shore, while the remainder of its surface is covered with numerous bowl-shaped depressions. These would, in all probability, have received the rising waters by ravines which enter the body of Ust-Urt from the low

[1] This is the height of the surface of the lake, which exists in the bed of the Western Manytsch, at its higher extremity, though the level of the banks, at the bifurcation of the Eastern and Western Manytsch channels, is more. M. Hommaire de Hell stated this height to be nearly ninety feet above the sea, which is not very incorrect, though perhaps slightly in excess of reality.

steppes upon its north and upon its south, and the aspect of the plateau would thus have been changed into that of the lake and marsh sprinkled highland whose traces remain to-day.

In this imaginary reconstruction of the Asiatic Mediterranean, the moment the rising waters reached a point at about two hundred and ten feet above the sea, and which is situated at the head of the now dry gulf Abougir, they would have entered into and filled up the basin of Lake Aral. A glance at the map will show this lowest point in the ground enclosing the basin to have been actually worn down from some greater height by an escape of the waters in a southerly direction, which took place in historical times. What such greater height may be precisely has not yet been observed, but the horizontal water-marks upon the sides of the Kashkanatao hills allow it to be stated, with sufficient correctness, that Lake Aral had, at a former date, its surface on a level of about sixty feet above what it was in 1874, *i.e.* at two hundred and nineteen feet nearly above the sea, which is approximately two hundred and twenty feet. Consequently, these figures will represent the maximum elevation of the lowest point in the transverse ridge north of Aral, by which a discharge of the Asiatic Mediterranean waters could have taken place towards the Frozen Ocean.

It cannot yet be certainly affirmed that, at a level of two hundred and twenty feet, a junction of the surfaces of the Caspian and Aral could have taken place through the gap at the southern end of the Mougodjar hills, by which the Emba steppes and the northern part of the Aral basin are connected; but the observations that have been made show the greatest probability of such a junction having occurred near the Tchagan stream, which rises in this locality. The passage of water, by a point situated in lat. 46° 25′ and long. (Greenwich) 58° 34′, whose height is two hundred and ten feet above sea level, would probably have drowned the still existent salt lakes and marshes, and the desiccated tracts covered with sand, which extend along the northern limits of Ust-Urt. These may in turn have supplied waters for other lakes which have now dried up, and which may also have been filled from ravines penetrating the cliff which is at once the eastern limit of Ust-Urt and the western one of the Aral basin. Such appears certainly to have been the case with the sand tracts and salt marshes called Barsakilmas, which a ravine named Kara-Umbet connects with the basin of Lake Aral in about lat. 43° 25′. The whole of the arguments which have been submitted indicate a very strong probability that the three basins of the Euxine, the Caspian, and the Aral could be connected again in the waterspread of an Asiatic Mediterranean, which would pass its surplus water to the north, merely by closing the Bosphorus to a height of two hundred and twenty feet.

Very strong evidence of the former existence of such an inland sea is to be found in the observations made by the naturalist Pallas in the basin of the Caspian. In his works will be found a map on which the ancient limits of this sea are traced as following a line directly northwards to the mouth of the Eruslan stream in the Volga along the right bank of this river. The line begins from the south-east angle, just below the summit of the Cholon Komyr plateau, upon the depressed channel of the Manytsch, by which the basins of the Euxine and Caspian are connected. From the mouth of the Eruslan the old shore passes to the east with a little northing, and skirts the southern slopes of the Obtschei Sirt hills, in the direction of Orenburg. It then turns south across the river Ural or Yaïk, below the junction of the river Ilek, and continues to some salt lakes on the Kirghiz steppe, beyond which Pallas does not appear to have followed it. Now, since the height of the south-east angle of the Cholon Komyr plateau is about two hundred and twenty-one feet above the sea, it follows that the Asiatic Mediterranean had a surface whose height may, with sufficient correctness, be stated at two hundred and twenty feet above sea-level.

If, now, it be supposed that with these aspects, the outlet of the Bosphorus were opened, the surface of the water in the Black Sea basin would more or less speedily fall to sea-level, while that of the Caspian would remain at the height of the ground which separates the two basins. This level would have been preserved, had the waters of the rivers which then supplied the Caspian been equal to the quantity lost from the surface by evaporation; but there is good reason for thinking that the Volga, previous to the Christian era, flowed into the Sea of Azof, which would probably—its level still being slightly above that of the Black Sea—have been then larger in extent than it now is. This is a circumstance that the descriptions left by ancient historians and geographers leave little room for doubting, and the change that has taken place may be explained by the constant enlargement of the Strait of Kertch. The level of the Caspian[1] would accordingly have fallen, since it would have been deprived of the waters of the great river now discharging into it; and that its surface did so fall is almost beyond question—since at Derbend, a very ancient city whose foundation is assigned to Alexander, masonry buildings are still to be seen at a depth of fifty feet in the water. Other circumstances even indicate that the Caspian could anciently be crossed dry-shod between latitudes 40° and 41°, *i.e.* its water surface was at least seven hundred feet below sea-level. Besides the former deprivation of the waters of the Volga, the Caspian certainly lost at a later date, first a part and then the whole of the stream of the Oxus, when that river changed its direction of flow northwards into Lake Aral.

As regards the basin of Aral, it is evident that, after the opening of the Bosphorus and its consequent separation from the Caspian, it would have been placed and would have remained in its present state of isolation, had not the quantity of water it received from its tributary rivers filled it up to over-

[1] It is now about eighty-four feet below sea-level.

flowing, and so caused its junction in another way with the Caspian.

Issuing from the more elevated basin of Lake Aral, the surplus water passing gently over the saturated marshy bed of the drained-off Asiatic Mediterranean, would have been prevented by dense growths of *arundo* (which already existed upon the shallower parts, or which would have immediately sprung up over the recently exposed surface), from acquiring on its passage to the lower levels a sufficient momentum to excavate channels for its passage. It would have diffused itself therefore laterally over the lower steppes to the north and to the west, and have also filled up the intersecting ravines and depressions of Ust-Urt. As regards a somewhat similar physical phenomenon, Humboldt has told us how he himself saw, in the Valley of Mexico, what a vast extent of ground could be covered with water by a flood of a few inches where the littoral of the Lake Tescusco is gently undulating or entirely level. And this point will receive further illustration on a future page, when the periodic inundation of a tract of country of many hundreds of square miles in extent upon the lower course of the Amúdarya, will be described as forming a rush-covered sea or mass of marshes and lakes; in spite of the fact that the slope of the ground is comparatively a great one, and that the flood is caused by a river running with a high velocity acquired in its descent from the elevated plateau of Pamir.

While such a discharge from Lake Aral existed it is clear that this basin would have been connected with the lower one of the Caspian, and that the whole extent of the land covered by such a united waterspread would have approached in shape to a triangle whose apex was on the north of Aral, and which had the west shore of the Caspian as its base. The central tract occupied by the water-surrounded plateau of Ust-Urt would have possessed in a lesser degree the aspects of a sea and would have been rather an archipelago. On the north, the depression extending to the Frozen Ocean, along the eastern foot of the Ural chain, would have been deluged. The quantity of water feeding these inundations would have varied with the height of the surface of Lake Aral, which would again have depended on the greater or less quantity diverted from its supplying rivers for the cultivation of the neighbouring country, or on the loss from evaporation in marshes, or on the radical changes in the flow of these rivers whose affluents may so have been deprived of a conducting-channel into the lake and partially or altogether lost in the sands. And since it may be certainly affirmed that all these three circumstances may or may not have acted at different times, singly or in combination one with another, or altogether, with different degrees of intensity, it is manifest that their consequences upon the body of water contained in the basin of Lake Aral would have varied very much, and that the escape of water from the basin of the Aral to that of the Caspian was an intermittent one. Thus nations at a distance, while having a general knowledge of the primitive form of the Aralo-Caspian Sea, and being even acquainted with some of its especial aspects, might have remained entirely ignorant of the conditions of its existence; though the populations living upon the banks of the more elevated basin could have accurately observed its actual isolation from the ground which was inundated by its overflow. In the following chapter historical evidence will be submitted which points to the absolute certainty that Lake Aral did overflow to the north in the manner which has been described, and formed besides such a junction with the Caspian as caused all ancient historians and geographers to regard the two basins so united as one and an identical sea; though the physical aspects were from time to time altered, in proportion to the volume of water which overflowed and flooded the country.[1]

[1] In this chapter, the present and prehistoric conditions of rainfall, &c., have been taken as identical. Enough has, however, been previously said to show that moisture has decreased enormously in the Aralo-Caspian region, and it is certain, therefore, that the quantity of water now draining down to the basin of the prehistoric Asiatic Mediterranean would be quite unequal to restore ancient physical aspects.

# CHAPTER XI.

## THE CASPIAN OF HISTORY.

Herodotus on the Caspian—Flooding of country between Lake Aral and the Caspian—Outlets of northern arm of Araxe—Strabo, Pomponius Mela, and Pliny on the Caspian—Arrian and Quintus Curtius —Chinese knowledge of the Caspian—The water of the Sea—Ammian Marcelline and Menander on Lake Aral—Flow of rivers from Lake Aral to the Caspian—Jenkinson's Gulf of Caspian—European maps showing a knowledge of Lake Aral—Olearius on shape of the Caspian—The Lake of Kitay of Herberstein, Jenkinson, &c.—European knowledge of isolation of Lake Aral.

THE first geographical authority on the Caspian is Herodotus, who affirmed the isolation of its basin in terms so explicit as to suggest some previous discussion on this point—perhaps, if a conjecture may be hazarded, in reference to the Argonautic expedition—but at any rate before his time, the ancient Caspian seems to have been connected by geographers and poets with the Ocean. The information of Herodotus does not warrant us in saying that the Caspian of his day included

Lake Aral, but the following considerations may perhaps indicate the probably recent aqueous connection of the two basins. What he says goes to show, that no northern discharge from Aral towards the Frozen Ocean was taking place in his day, and it may therefore be deduced that the rivers of Turkestan contributed, at that time, but a limited volume of water into their recipient. The description given by Heeren of the traffic between Olbia, the Greek mart at the mouth of the Borysthenes, and Central Asia, that is grounded upon passages in the fourth book of Herodotus, affords some indication of the truth of the statement which has just been made. For this flourishing trade points to the civilised and prosperous condition of these regions; and such a social state would imply a large population, demanding an extensive cultivation of the soil, that could only have been carried on by the diversion of much water from the rivers feeding Lake Aral, thus causing a fall in the level of the waterspread of that basin.

However, in speaking of the country of the Budini, the historian describes it as thickly wooded, and containing in one part a broad deep lake, that was surrounded by marshy ground having reeds growing in it, where otters and beavers were caught, besides another sort of square-faced animal. Now if, as has been generally supposed, these animals were seals, the lake in question, in order to allow of their presence in sufficient numbers to be hunted, must have been situated in the Aralo-Caspian depression, which extends northwards into the Frozen Ocean; and if this lake of the Budini did not occupy the basin of Aral itself it would have been a sheet of water due to that overflow from Aral in a northerly direction, which we gather from Herodotus' description must have been in temporary abeyance.

And here the questions are suggested: who the blue-eyed, red-haired Budini could have been, and where was their town Gelonus, that was surrounded by a lofty wooden wall? Can Gelonus have been the ancient forerunner of the great centre of commerce, Gorgándj,[1] which is described by the early Arabians, and from which Ibn Haukal, writing in the tenth century, tells us the caravans used, in *old times*, to go to the land of the Khazars? And since the Turanians have usually dark hair and eyes, is it

[1] Al Djordjáníja, whose place was subsequently taken by the present Kunya Urgenj.

possible the Budini may have been of the blood of those early Aryans who worked out their civilisation along the banks of the rivers of Western Asia? In such a case, the fugitive Greeks from the marts of the Euxine would have found an appropriate asylum in Gelonus, as well as a ready field for the reception of their religious rites, their customs and language, which Herodotus tells us were all more or less in vogue in this town.

Heeren has traced the early trade route of the Greeks into Central Asia as passing from Olbia by Astrakhan, and thence to the north-east, and in this latter portion it almost exactly follows the line drawn by Pallas on his map as the former shore of the Caspian. It is evident, Heeren says, that this line was not the shortest one, but it is the most southerly that the description given by Herodotus will allow; and, as an explanation of the *détour*, he goes on to suggest that the fear of brigands, who frequented the shorter and more desert *route*, may have rendered it necessary to be made. According to the explanation which is here being developed, it seems more probable that the drowned state of the country between the Caspian and Aral (notwithstanding the temporary cessation of the overflow from the latter sea) may have been the reason why caravans did not travel across it; for Istakhrí and Ibn Haukal, the Arabian geographers, writing in the tenth century, tell us that Ust Urt and the Mangishlak peninsula, had only been lately inhabited at that date, information which indicates some practical difficulty having attended the former occupation of this country by the nomadic tribes of Asia, which may very well have been its marshy state.

That the terminal basin of the Jaxartes possessed at that time a low level is, perhaps, also shown by the description given by Herodotus of the northern arms of the Araxe (or Oxus), which were stated to end in a region of lakes and swamps, inhabited by people who dressed in seal-skins; and these outlets, according to the argument which has already been used, would have been in the great Aralo-Caspian depression. This region of lakes and swamps must certainly have been to the south of the transverse watershed which extends along latitude 51°; yet the previous existence of the Asiatic Mediterranean, and the intermittent overflow to the north of Lake Aral which succeeded it, would have permitted the ascent of seals from the

Northern Ocean, and so account for their presence to-day in the Caspian, though they have been extinct for half a century in Lake Aral, which has been the seat of the greatest desiccation of the Aralo-Caspian region.

The confusion that was wilfully made by the geographers of Alexander's day, between the Tanais (Jaxartes) of Asia, which was the farthest limit attained by the arms of the conquering Macedonian, and the Tanais (Don) of Europe, whose mouth was in the Palus Mœotis, may have been suggested by the swampy shores of the basin, which received the first of these two rivers. It results from what has already been said, that such marshes would have been more or less continuous in a westerly direction, to the northern part of the Caspian, which ancient writers tell us received a supply of water from the Palus Mœotis. At that date, this latter body of water also covered a larger extent of land than it now does. Its level would consequently have been higher, and surplus water could have passed from its swampy edges by the channel of the Manytsch into the Caspian basin, while reason exists for thinking that this took place, even after the Christian era. The error that was designed to flatter the vainglory of Alexander, by extending his conquests over the distant steppes which were occupied by the Asiatic Scyths, as far westerly as the Tanais of Europe, was thus assisted by the similar physical aspects of the whole country intervening between the Aral basin and the Palus Mœotis.

From the earliest historical times there have thus been indications of the great extension of the Caspian to the east, but it is only in the first century of the Christian era that the statements of geographers afford firmer ground, on which more certain conclusions can be founded, regarding the physical aspects of the sea. About this epoch there can be scarcely any doubt of the state of the basin of Aral, as the following considerations will suffice to show. Strabo tells us that in observing how the Caspian advances in a southerly direction from the Ocean, it may be said that it properly forms a gulf, since, from the narrowness it possesses at its entrance, it widens out as it penetrates the land. This entry of the ancient Caspian from the Ocean into Asia, by a long and narrow canal, is also described by Pomponius Mela and Pliny, who state that scarcely had the waters begun to assume breadth than they curved round on both sides in a crescent, and taking the form, as it were, of a sickle, formed three gulfs. The first gulf, which was opposite to the mouth of the canal mentioned, was the Hyrcanian Gulf, that to the left was the Scythic Gulf, while that to the right was the one which was, properly speaking, the Caspian. It would be difficult to describe in plainer language the appearances which would be presented by a lake occupying the position of Aral, and overflowing to the north, as well as westwards towards the Caspian, as has been shown. The escape to the north would imply that the volumes of the rivers discharging into Aral had, since the time of Herodotus, increased to such an extent as made his description of the isolation of the Caspian Sea no longer applicable. It may, however, be gathered from Pliny, that even previously to the Christian era, Lake Aral was overflowing westwards, for he states that nearly the whole of the northern coast of the Caspian was explored by row-boats, under the reigns of Antiochus and of Seleucus, while Arrian expressly mentions that the limits of the sea were unknown in Alexander's day, though Quintus Curtius tells us that it covered a great extent of country with water of little depth on its northern side.

The navigation of such a sea as was formed by the overflow of Aral towards the Caspian would probably have been more easy in passing to the lower basin on the west than in ascending to the higher one on the east; not because there was necessarily any very strong current, but rather because the channels leading through the mass of rushes, which have been indicated as covering this inundated country, would have been tortuous. We read, in reference to this, that when the Chinese (who knew of no difference between the Aral and Caspian) advanced, under the general Pan-tchao, to the Western Sea, in the first century of the Christian era, they found its shores covered by extensive marshes; and, wishing to push their conquests still farther to the west, they had the intention of crossing the sea, but abandoned the idea for the following reason: The Tajiks (i.e. the indigenous inhabitants) of those localities told Pan-tchao that the navigation he was about to undertake was very dangerous, for according to their statements it required two months with a favourable wind to cross the sea, but to get back (i.e. from west to east) it required two years; so that those who navigated it and wished to make the return trip, were in the habit of taking three years' provisions with them. It appears, then, from this passage, that the Caspian was navigated previous to the first century of the Christian era, from the direction of Aral westwards, a circumstance which indicates the overflow supposed to have existed.

The Western Sea of the Chinese would naturally be the Eastern Sea of the Greeks and Romans; and in Pliny there is a passage relating to the quality of its waters. He says the waters of the Eastern Sea were sweet—a statement resting on observations made in the time of Alexander and on the authority of Varro, who saw some of it brought to Pompey during the Mithridatic war, i.e. the first century B.C. Without doubt, adds Pliny, 'it is the enormous mass of water brought by the tributary rivers that neutralises the saltness,' which he supposed it ought to have possessed from its connection with the Ocean. The passage confirms in a very remarkable manner the hydraulic conditions which were assumed as existent and as being necessary for the filling up of, and for the escape of water from, the Aral basin. Pliny also tells us that all that Sea was covered with islands, in which notice we have, perhaps, an indication of the lake and marsh-covered surface of

Ust Urt. In face of the statements made by Strabo, Pomponius Mela, and Pliny regarding the physical aspects of the Caspian about their epoch, it is only possible to reconcile Ptolemy's information of a century later regarding its isolation from the Ocean, by supposing a subsequent decrease of water in Lake Aral.

Such an aspect is indicated by the passage in Ammian Marcelline (fourth century) concerning the marsh which received the rivers Araxetes and Dymas, as well as by the notice in the relation by Menander (in the sixth century) of the embassy of Zemark to the Khan of the Turks. Both of these point very strongly to the low level possessed by Aral at those dates, and consequently to the continually growing desiccation of the country between it and the Caspian. Still, as Humboldt remarks, the Argonautic and Alexandrian tradition regarding the connection of the Caspian with the Northern Ocean, was generally adhered to by European geographers, for many centuries subsequent to Ptolemy's day.

We now enter upon the records of the Arabian geographers, who were perfectly acquainted with the physical aspects of the Aralo-Caspian depression. M. Ali Suavi[1] states that in the seventh century the

[1] In 'Bulletin de la Soc. Geo. Paris' for November 1873.

Arabians who conquered Central Asia were acquainted with Aral. Be this as it may, in the ninth century the two writers, Ibn Khordadbeh and Jakubí, nevertheless considered the Aral and the Caspian as forming a single sheet of water, just as many geographers did down to very recent times. In the tenth century, however, Ibn Dosteh gives the Lake of Khwarezm a circumference of three hundred and twenty miles, and says its western shore is enclosed by mountains called Siakoh, which are evidently Ust Urt. Masúdí also informs us that from the Lake great rivers ran into the Caspian Sea, so that it appears the overflow, which began some twenty-five centuries previously, after the rupture of the Bosphorus, had gradually resolved itself at this date into decided streams. Still later we find that the overflow from Aral ceased altogether; for, though the route from the Volga to Central Asia seems to have followed in the thirteenth century the present northerly line, in the fourteenth century it also passed from Sarai-chik on the lower Yaïk, in a south-westerly direction, over the Ust Urt plateau, to Kunya Urgenj; and must consequently have crossed extensive areas of country which were formerly flooded. Aral, we are, indeed, told by a Persian writer whose MS. is in

Sir Henry Rawlinson's possession, had at this time dried up, in consequence of a change of flow of the Oxus to the Caspian, as well as of the passage of the waters of the Syr into that river.

The English traveller Jenkinson found no overflow taking place from Aral in 1559, though an expression he uses in the account of the travels he made in these regions, is a very strong confirmation of the truth of the idea that the ancient Caspian was composed of its present basin joined to that of Aral by the overflow of this latter sea. After travelling for twenty days across waterless deserts in an easterly direction from the shores of the Caspian, Jenkinson arrived at what he calls 'a gulfe of the Caspian sea againe,' where he found the water very 'freshe and swete.' Now this place is, doubtless (as will be seen on a future page), the Lake Sarakamish, which is situated in a bay or indentation lying in the Tchink or southern limiting cliff of Ust Urt, and which would have actually been a gulf of the Caspian at a time when a sufficient overflow of water from Aral united the two seas in the way under explanation.

A knowledge of the isolation of the Aral and Caspian basins, which was accurately recorded by the Arabian geographers so far back as the tenth century, must have passed, by some means, to the West; for, though the name of Aral does not appear to have been known to Europeans, the map of the Venetian Marino Sanuto in 1325 shows the *Mare Caspis, Yrcanum*, or *de Sara*; and beyond this, at some distance to the east, is a small lake without a name, into which Gyon (Oxus) flows, while between these is a second *Mare Caspis,* full of islands, closely shut in on the west by the *Montes Caspii.* From this lake full of islands runs a river through these mountains, into the larger *Mare Caspis* or *Yrcanum* on the west. Towards the end of this century we find Cardinal Pierre d'Ailly expressing an opinion that there were two Caspian seas, one entirely surrounded by land, and another which communicated with the Northern Ocean; and both of the circumstances mentioned point to increasing European knowledge on the growing desiccation of these regions. Still the error, by which the greatest dimension of the Caspian was shown extending east and west, was not corrected until the seventeenth century, and even continued to exist on many respectable European maps till so late as the eighteenth century. On the return of Adam Olearius, the Swedish Ambassador, from a visit to Russia in 1636 he wrote a book, in which he says 'that the length of the Caspian does not extend from Levant to Ponent, as one sees it drawn on all maps, even in those which have been published since the first printing of this relation, when I condemned the error; but that it must be taken from north to south, and that it is its breadth, on the contrary, which stretches from Ponent to Levant.' He then gives the authority on which he makes the statement, and continues: 'Upon this information all the geographical maps must be corrected, though the opinion sought to be established is a new one, and directly opposed to that which has been received so many ages since.'

As to the overflow from the basin of Lake Aral to the north, the Baron de Herberstein, in the sixteenth century, and Olearius, in the seventeenth century, both state the Obi to take its rise in the Lake of Kitay; while Peter Appian, at the former date, still considered the Caspian to be a gulf of the Northern Ocean. The above source is also shown for the river Obi in the map of Ortelius, that illustrates Anthony Jenkinson's journey to Bokhara. This lake of Kitay, into which Jenkinson's map shows the Syr to have discharged its waters, seems to have been that basin on the north of the present waterspread of Aral, which is occupied by the semi-desiccated salt lakes Chalkar-tengis and Sary Kupa, the recipients of the Irgeez and Tourgai rivers. Shumsuddín Dimashkí, an Arabian writer of the end of the thirteenth century, states that the Syr joined the Amú, and that between the point of junction and the outlet was ten days' journey. The junction of the two rivers seems thus to have been in the neighbourhood of the present south shore of Aral; and the Lake of Kitay, in the basin of Chalkar-tengis, might consequently have been the receiving basin mentioned by Shumsuddín in this passage, which would indicate the present sea of Aral to have been at such a time in a desiccated state—the circumstance that is mentioned by the Persian writer of the early part of the fifteenth century. It remains, however, to be determined by actual observation whether the height of the ground surrounding the basin of Chalkar-tengis will allow of the conditions of northerly overflow, and of partial isolation from the actual basin of Aral, which are indicated by the preceding information, as well as by that which follows.

There is a very curious letter which will be

found in the travels of Bergeron, published at the Hague in the eighteenth century, and which was written to the geographer Mercator at the end of the sixteenth century, affording strong evidence that the basin north of Aral had been in a recent state of overflow. This letter gives the story related by a Flemish soldier who had been a prisoner in Russia, and who had visited the river Obi on trading expeditions. There he was told that the river could be ascended to the Lake of Kitay, from which it flowed, and that upon the surface of the lake boats were met carrying merchandise and manned by black men, which reached this body of water by the great river Ardok, i. e. the Oxus or Amú-darya. We have in these circumstances a very striking example of the rapidity with which the basin of Aral was soon after filled up, from its previous state of desiccation, by the change of flow of the Oxus, which in the sixteenth century abandoned its Caspian outlet for one in Lake Aral.

The actual isolation of the two basins, that has been well known in recent times, has given rise to many attempts to explain the historical facts which had been recorded regarding the Aralo-Caspian Sea. The present theory has been grounded upon the observations made by Russian officers sent by the Imperial Geographical Society to these localities in 1874. These researches afford the strongest presumption for affirming that no such movement or upheaval as has generally been supposed, has taken place in the Aralo-Caspian region since very remote times. The theory of an intermittent overflow from the basin of Lake Aral, which has been developed in the foregoing pages, and which is the logical consequence of the observations mentioned, is submitted to the consideration of the reader, as very fairly reconciling many conflicting historical statements, as well as agreeing with the physical conditions of the localities described in the past and existent at the present day, as far as such are certainly known, besides agreeing with others but partially known.

# CHAPTER XII.

## THE CENTRAL OUTLET OF THE AMÚ-DARYA.

The south shore of Lake Aral—Saltness and colour of water—Earthy matter carried down by river—False bottom of the Lake—The three lower arms of the river—Shallowing of south shore—The Kichkine mouth—Process of formation of its delta—The banks of the Kichkine-darya—Karakalpak pasturages and cattle—Akkala, the 'white fort'—Its engagement with the 'Samarcand' in 1873—Mooring of steamers at night on river's bank—The Ulkun-darya—The Karakalpaks of the Lower Amú.

THE southern shore of Lake Aral, along which the mouths of the Amú-darya are situated, has a low monotonous coast-line of about eighty miles in length, edged with a tolerably high and thick growth of rushes. These are frequented by numerous flights of pelicans and other water-fowl which form, with their enemies the eagles and the hawks, the distinguishing ornithological features of these regions. Above the line of green rushes at the water's edge rise low ridges of dun-coloured sand, and through an occasional gap between them are seen the beehive-looking *kibitkas* of the Karakalpak encampments, from which a few stray camels near the glassy water loom large through the heated atmosphere. At about three miles' distance from the strand the saltness of the waters of the Lake perceptibly diminishes, and the change continues until, on reaching the Kichkine mouth of the Amú, the bitter taste has entirely disappeared. Along the same distance, during the flood season of the river, the clearness of the lake waters becomes clouded, and their colour changes close to the shore to a yellowish brown; but during the remaining two-thirds of the year little or no earthy matter is brought down by the Amú waters, which then reach their outlets in a clear condition.

The large quantity of earth carried in suspension by the Oxus is one of its striking and most important characteristics, and was noticed by Polybius two centuries before Christ, and sixteen centuries ago by Quintus Curtius, in his history of Alexander the Great's Asiatic expedition. It may, perhaps, be thought that deposits made by the river in the basin of Aral must have a very appreciable effect in filling it up and so raising the level of its contents; and though this is, no doubt, the case in a limited degree, the quantity of earthy matter brought into the Lake is practically small, from the circumstance that it is only during the flood season, when but one half of the stream reaches an outlet, that matter in suspension is conveyed by the river. Of the earthy burden carried by the waters reaching Aral during this season of the year, much is left behind in the lakes and swamps upon the lower courses of the Amú, while that due to the other half of the stream which is diverted for irrigation purposes in Khiva, is deposited on the surface of the Khwarezmian plain. Such earth as enters the Lake is exceedingly fine, and, settling down gradually, remains upon the vegetable growths of the bottom, and so forms a false bed, through which a stiff bar can be thrust or a sounding lead will pass. The weed spoken of does not appear to be a true seaweed, for unsuccessful attempts have been made to utilise it as a stuffing for sailors' mattresses, with the invariable result of resolving it speedily into a powder.

Rather more than eighty miles directly south of Lake Aral is situated the Khivan town of Khodjeili, upon the left bank of the Amú, which here begins to separate into the various arms by which its waters reach their three outlets. Placing oneself at this place and facing the north, one has on the right the Yany-Su, or 'new water,' the most recently formed arm of the river, which flows away in a north-easterly direction to its mouth, but which midway in its course passes through an extensive shallow lake called Dowkara. The older portion of the Amú flows to the left, or north-westwards, from Khodjeili down to the town of Kungrad, where it turns due north and enters Lake Aral, after a direct course of about forty miles, by the arm called Taldyk or 'straight.' In its flow between Khodjeili and Kungrad the river gives off several channels from its right bank towards Aral, and the streams of these fill a depressed tract of lakes and marshes, which are drained off below by the Ulkun or 'great' arm of the Amú-darya; whose upper course leaves Taldyk at Kungrad, and flows in a north-easterly direction to the point where it receives the drainage spoken of. The Ulkun disembogues by several small branches, of which one is the Kichkine.

Each of the three arms of the river to Lake Aral has its own proper *delta*, formed of the earthy matter brought down by their streams; and, as might be expected, the older or western mouth of Taldyk has the most extensively developed one. In the neighbourhood of this the strait situated between the mainland and the island of Tokmak Atta is gradually shallowing, and the cattle of the Karakalpaks now swim over to the island without difficulty. A perceptible decrease of the depth of the Lake was also noticed between 1848 and 1858, and again in 1874, near the eastern mouth of the river, and some of this shallowing is, no doubt, due to earth carried down by the Amú, but must in a still greater degree be referred to the gradual fall of the level of the water in the Aral basin. Boutakoff found in 1848 three feet depth of water at the entrance of the Abougir Gulf, which had only eighteen inches at the same spot in 1858, and is now entirely dry, as has been mentioned among the evidence adduced of the fall taking place during recent years in the level of Lake Aral. From the years 1848–49 when the Russian officer just named first explored the lower Amú-darya, down to 1873, the year of the Khivan campaign, the river has been little visited; except in 1858, when a second examination was made by a flotilla steamer.

A passage of about fifteen hours under easy steam

brings the Russian vessels from Kos Aral to the mouths of the central arm of the Amú, where a small pilot establishment is kept up for buoying out the deep channel on the approach of a steamer. This outlet has rather more water and is broader than the single navigable arm of the Syr-darya, though in all probability both entry and exit by it would be difficult much after the cessation of the river floods. From two hundred yards of breadth and seven feet of *minimum* depth possessed by this, the Kichkine mouth of the Ulkun, the channel a quarter of a mile inside decreases to barely a hundred yards in breadth, though the depth increases to twenty and even to thirty feet a little way farther up the stream, where the breadth also again increases to more than two hundred yards.

The lower quarter of a mile of the channel flows directly out to Lake Aral from the river which runs parallel to the coast from the east, and at the elbow another small channel continues to the west, and discharges into the Lake by several smaller outlets, into which it divides in that direction. Seven or eight miles up the river, from the same elbow, is another large outlet which is the original mouth of the Ulkun-darya arm of the Amú. The peculiar form of this *delta* is due to the deposition of sand by the stream at its different mouths, acting in conjunction with the fall taking place in the surface of Lake Aral. As the volume of the flood and the velocity of the Amú decreases sand is deposited, and adds to the bar at its mouth on the *inside*; and as the Lake, deprived of its supply of water, quickly falls in level from surface evaporation, sand is blown and washed up and added to the bar on the *outside*. At the epoch of the flood of the following year the stream, embarrassed by these deposits at its mouth, is unable to get rid of its waters with sufficient rapidity, and gradually extends a channel along a line parallel to the shore, until a favourable point occurs for the formation of a fresh outlet. A succession of similar phenomena seems traceable at the south-eastern corner of Aral, where a number of generally parallel outlets appear to have been formed by the Jánídarya arm of the Syr when the Lake had a higher level than at present.

The banks of the Kichkine channel are very little elevated above the surface of the water, and the edges are lined with the usual dense beds of tall rushes, behind which is a low, scrubby, thorny jungle.

The ground, however, is generally open, after ascending a short way, and consists of fine wide pastures, occupied by the cattle of the Karakalpaks, whose encampments were passed at every mile or two, as the 'Samarcand' passed along the lower twelve miles of this arm of the river. The number of these people seen was, perhaps, four or five thousand souls, or one eighth nearly of the whole tribe; and whether they or their cattle looked the more astonished as the 'Samarcand' steamed slowly by, with two transport barges in tow, it would be difficult to say. One enterprising bullock nearly found a watery grave, by plunging into the water and swimming after the monster with the smoking funnel, whose strange appearance evidently piqued the beast's curiosity. Other animals were more generally timid upon the steamer's approach, and many hundreds set off at a gallop and continued their flight across the pastures far into the distance. Sometimes on the banks were small plots of cultivated ground, and from the river a few irrigation cuts ran towards the south-west, on which horizon a solitary tree or two showed large against the sky. The Karakalpaks raise a small quantity of cereals, and the crops are stored in *zamoffkas*, or mud-walled enclosures, which also afford some shelter for their *kibitkas* in winter, and are often seen on the banks of the lower Amú. Besides these structures there are a few ruined mud forts, which formerly belonged to Khiva, and on whose walls are invariably seen perched some of the gigantic eagles that are special to this part of the world.

The 'Samarcand' had entered the Kichkine mouth of the Amú late in the afternoon, and at sunset she passed the Ulkun mouth, while some half-an-hour after she reached Ak-kala or the 'White Fort.' This is situated on the right bank, near the first elbow the river makes to the west, and its crenelated mud walls form a square of about one hundred yards a side, which has circular flanking towers at the angles, and is surrounded by a ditch. The fort-walls are built of *pisé*, and surmount a rampart faced with a sort of cyclopean work, which consists of large blocks of hardened mud, fitted together in irregularly-shaped pieces, that are moulded or cut from the bed of the river.

The walls have a command of nearly twenty feet over the exterior of the fort; and had it been properly defended by the Khivans in 1873 the place might have been a hard nut to crack for the little popguns of the Aral flotilla. In the month of April of that year the 'Samarcand' was steaming slowly, but somewhat too confidently, past Ak-kala, which was occupied at the moment in some strength by the Uzbeks, and armed with two guns, that were discharged, and killed or wounded some half-dozen of the steamer's crew. The 'Samarcand' replied with a few rounds, and caused the place to be quickly evacuated; but, having passed on without first possessing herself of the armament of the place, the Khivans subsequently returned, and removed their guns before the Russians paid it another visit. At the present day this fort, with its adjacent enclosed village, is a mass of deserted ruins.

At nightfall the Russian steamers navigating the Turkestan rivers generally bring up alongside the bank, instead of anchoring in the stream; for the double reason that repairs of some sort have usually to be executed, and that the fuel bunkers require replenishing with *saksaoul*, which is carried on the transport barges in tow. In this way the soldiers and sailors are not seldom on fatigue half the night on their trip up a river, and cheerfully enough they perform the work, all disagreeable as it is made by the swarms of mosquitoes which fill the air. On such occasions there can be few more picturesque sights than the groups of men carrying the logs of firewood to the steamer, and lit up by the flickering flames of a roaring fire on the bank; where the artificers are making good some broken piece of iron-work belonging to the paddle-wheels or machinery.

At sunrise next day the 'Samarcand' had steam up, and, after twenty-five miles' further ascent of the river, gained the central tract of the lakes and marshes of the lower Amú; but just before reaching these she left the Ulkun, which flows down from the Taldyk arm, at Kungrad on the south-west. Above Ak-kala the right bank of the Ulkun branch is swampy, and across these marshes runs a now obsolete channel of the river, called Kazzak-darya. The opposite bank continues to be occupied by the *aouls* of the Karakalpaks, who have constructed small embankments on the lower parts, to preserve their lands from being flooded. This country, cut up as it is by lakes and swamps, is evidently unfavourable to the old nomadic proclivities of these people, for, though a single family, with their household goods, *kibitkas*, and a few sheep and cows, may move about the lower Amú in *caïques*, means of transit are, of course, wanting for the migrations of numbers of people with numerous flocks and herds. The tribe, having no ground suitable for camels, possesses but few of them, and has given up the breeding of these animals entirely, while there is but little proper ground also for horses and other cattle; so before long it seems pretty certain that the Karakalpaks will become permanently attached to the soil, which, with moderate labour, will give them a fair return. In the 'struggle for existence' at present engaging the energies of these people, a favourable opportunity occurs for an experiment on a large scale upon the reclamation of a nomadic race.

# CHAPTER XIII.

## THE LOWER AMÚ-DARYA.

The Amú-darya has no delta—The river between Khodjeili and Kungrad—The Abougir Gulf—Kungrad—Taldyk—The Ulkun-darya—Tumalaktao—Burying on hills, an important characteristic in the Oxus countries—Cultivation near the Ulkun—Its deterioration as a navigable arm—Tendency of Amú to pass by eastern arm—Kuwán-jerma—Kigailie—Town of Cimbye—Bi-weekly fair—Cultivation—Fields and Homesteads—Description of a homestead of the Oxus countries.

THE triangular tract of country which has the Khivan town of Khodjeili for its apex, and whose three sides are formed by the eastern and western arms of the river and by the south shore of Lake Aral, is frequently described by the misleading term, the Delta of the Amú-darya. It is true that nearly all large rivers divide towards their outlets into several branches, which are comprised more or less roughly in a triangular space; but such a Δ, or delta, is always formed by the earthy deposits carried down by the waters themselves, the slope of the ground from the apex of the Δ to the base being a very gentle one. In the Amú-darya this is not the case, since the point where the arms separate is sixty feet above the level of the Lake; whose southern shores have consequently a fall many times steeper than that usually possessed by the true delta of a river. As will be presently seen, the Amú reached Lake Aral, about two and a half centuries ago, either by the western arm, flowing directly north past Kungrad, or by that branch of it, running to the north-east from the same place, which is known as the Ulkun-darya; while it is, perhaps, not more than a hundred years ago since the eastern or Yany-Su arm effected an entry into the Lake. As regards the central area of lakes and swamps, which is situated between the eastern and western arms of the river, it is certain that this has been if not entirely formed, at least very considerably added to, by the channels which have opened from the right bank of the Amú, in its course between Khodjeili and Kungrad, during the past thirty years. So far, then, from the different lower branches of the river having made the ground through which they flow, with the earth their waters have deposited, it is certain that they have cut their channels during very recent times through the country that previously existed. The distinction is an important one, and if the somewhat extended examination of the physical phenomena which characterise the flow of the Amú, and which will now be entered upon, be found tedious, it may deserve study, since it will assist in elucidating the causes of the changes that have taken place in the flow of the Oxus, during the past twenty-four centuries. In this interval of time, history tells us, it first entered the Caspian, then changed its outlet into Lake Aral, to return once more to the Caspian, and finally to transfer its flow again into Aral by its present channels.

On reaching Khodjeili from the south, the Amú, after detaching its eastern arm, makes a decided turn to the west, and flows in a general north-west direction in a large volume of water, about four hundred yards broad. In the flood season this has a current of five miles an hour, and is of a reddish-brown colour, from the quantity of earthy matter it carries in suspension. In the portion of its course, down to Kungrad, the deep channel of the river travels from bank to bank, and is in consequence always changing as the shores are cut away. On the left or Khivan bank the country is cultivated, and houses are seen amidst the groves of trees, which form a leafy screen growing parallel to the river at no great distance. On the opposite or Russian bank is a low scrub jungle of tamarisk, jidda, and acacia ; and through this pass in a northerly direction, several branches given off by the river, and carrying their streams down to the central group of lakes and marshes. Near the upper end of the easterly one of these branches are situated the ruins of Tuk-kala, an ancient fort there will be occasion to refer to on a future page.

In former days the lower Amú, before reaching Kungrad, detached several channels from its left bank in a westerly direction, whose waters emptied into the now dry Abougir Gulf, which runs north and south, from the south-west corner of Aral, beneath the cliff of the Ust Urt plateau known as the Tchink. The gulf is about eighty miles long and ten miles broad, and also received some years ago the waters of the Loudon canal, which left the Amú-darya at a point some twenty miles above Khodjeili, and had its outlet near the head of Abougir ; but this stream was cut off, owing to the Khan of Khiva having erected a dam across its upper end in 1857.

This circumstance, in conjunction with the fall that has taken place in the level of Lake Aral, has drained the bed of the Abougir Gulf, and has allowed of its cultivation by a branch of the Yomut Turkomans.

On the left bank of Taldyk, about forty miles above its mouth in Lake Aral, stands the Khivan town of Kungrad, a place that has been formerly visited by Europeans, of whom Abbott is among the number, and has been more recently described by Vámbéry, in his Central Asian sketches. At the present day it is said by the Russian traders who know it, to consist of little more than a mass of *pisé* ruins; which border narrow and malodorous streets, where numerous *kibitkas* are pitched, since they are preferred as residences to the houses. The population of the town, stated twenty years ago to be six thousand souls, has fallen in numbers to two thousand, and is composed of Sarts, Uzbeks, Turkomans, Karakalpaks, and Kirghiz, who are governed by an old personage of the last-named tribe in Khivan employ. The women of Kungrad, of whom some probably ill-founded scandal has been repeated, are reported to be good-looking. The town itself may have derived its name from the Turanian tribe which is now assimilated with the Uzbeks, and which in former times usually supplied the harems of the great conquering Mongol-Turk families.

From Kungrad, Taldyk flows down to Aral between elevated banks (more so on the right hand), which are cut in an indurated clay soil. In this part the Amú was first examined by Boutakoff in 1849, when three feet of water was found on the bar, which ten years afterwards is stated to have advanced northwards more than half-a-mile, and to have had one-and-a-half feet of water only upon it. The mouth of Taldyk at the present day is almost entirely closed, and the whole of the lower course of this arm of the river is being gradually filled up with fine mud, which is deposited by the limited stream of water still passing through it to an outlet during the flood season. It thus appears that less than three centuries will probably elapse between the first opening and the ultimate closing of this mouth of the Amú-darya, from the operation of phenomena which characterise the river's flow.

To the north-east from Kungrad the Ulkun-darya flows out of Taldyk a distance of twenty-five miles before it receives the drainage of the central lakes;

after which it turns due north in the direction of Akkala, as has been described. On its upper course, along the distance mentioned, though the left bank is firm and elevated, the right bank of the Ulkun during high flood is scarcely distinguishable from the central swamps, which at this season extend themselves westwards. Immediately below Kungrad the stream contours and cuts into the foot of the isolated hill of Tumalaktao, and exposes a bed of soft sandstone, in which oyster-shells and sharks' teeth are interspersed, and cemented together with iron, giving brilliant red and yellow colours to the whole mass. Tumalaktao has an elevation of from one hundred to one hundred and fifty feet, and is a bare, rounded clay hill, strewn over with sand and small fragments of ferruginous sandstone, and sprinkled with a few prickly plants. There is a semi-ruinous mud village and some cultivated ground, besides a large Karakalpak *aoul* at the foot of Tumalaktao, whose summit is used by the nomads as a cemetery. This custom of burying the dead on hill-tops only, is invariable in the countries that are watered by the lower Oxus. It is necessary, in order to preserve human remains from disturbance by inundations, and is an indication of the physical phenomena which have attended the flow of the river as well as of the varying surface of Lake Aral since times of the remotest antiquity.

On the banks of the Ulkun, in the neighbourhood of Kungrad, there is a good deal of agriculture, and the irrigating water is raised from the stream with Persian wheels worked by horses. The cultivated fields are either melon-beds or small patches where wheat, barley, and oats, all mixed up together, are grown; or of lucerne, besides a pea which is crushed for its oil and then used as food for cattle. On the moister ground are rice-fields, where the crops are guarded from numberless wild ducks by yelling Karakalpak urchins. The left bank of this branch is generally covered lower down its course with excellent pasturages, alternating with patches of low thick jungle, where hares, gazelles, and pheasants swarm.

The breadth of the Ulkun, down to where it receives the drainage of the central swamps is less than one hundred yards, and its *maximum* depth is never more than six feet. It was ascended by Boutakoff in his early explorations of the Amú, when it carried a much larger volume of water than it now does, and as a navigable stream it has been wholly destroyed by some half-a-dozen dams, which the Uzbeks constructed across it in order to prevent a second ascent of the Russian steamer whose appearance so thoroughly frightened them in 1848. The obstructions were broken down by the Russians in 1873, with a view of assisting the operations of the Aral flotilla during the Khivan campaign; but no advantage resulted, for the dams had effectually done the work of filling up the channel with deposit, and had thoroughly established the tendency of the Amú to abandon its western for its eastern outlets.

The eastern arm of the river branches off immediately below Khodjeili, under the name of Kuwán-jerma, or the 'new cut;' and towards its commencement its very slightly elevated banks are clothed with a tolerably thick jungle, besides some few trees, such as poplars, dwarf elms, and willows, which owe their origin, probably, to planting. Such vegetation, however, only occurs on the firmer ground, for, on the lower levels, which are subject to inundation, there is a heavy growth of tamarisk bushes, whose delicate pink flowers contrast beautifully with the sombre green foliage.

Fifteen miles below the head of Kuwán-jerma the large canal called Kigailee is detached to water the country on the left bank, and also to serve as a line of communication with the town of Cimbye. This is situated midway between Lake Dowkara, which Kuwán-jerma enters, and the Kashkanatao hills, which border the central Amú swamps on their east. From the town the country falls in the Dowkara direction, and during the flood season is marshy and inundated by streams of water which leave the eastern arm of the river below Kigailee.

Cimbye, the second town of the lower Amú-darya, has a population of about twelve hundred souls, and contains a bazaar where Central Asian silks, Russian calicoes, embroidered skull-caps, sheep-skin *shubas*,[1] and painted wooden saddles are the chief articles of commerce. The bazaar shops are of the usual Oriental type, and recall those of the meanest at Suez. An open-air market is held outside the town twice a week, when two or three thousand people come in from the neighbouring country to make purchases or to sell cattle and agricultural produce, and as every man, woman, and child rides either horse, or ass, or camel, the scene is an animated and amusing one. Each vendor in the market pays a *percentage* of the price of the articles sold to the

[1] Fur pelisses.

Russian Government, represented in the person of a ruddy-faced Kirghiz from Kokand, who is the chief of the place, and whose features have a very marked Mongol type. This individual, clothed in gorgeous silks, and girt with a silver belt, sits cross-legged under an awning in the centre of the market-square, which is filled with a crowd of bawling human beings, and of screaming, neighing, and kicking animals. Drinking his tea, and smoking his cigarettes *à la Russe,* he presides at the receipt of custom, and converses affably the while with the more notable among the visitors to the fair.

The Kigailee canal passes through the town and fertilises many orchards and fields in the neighbourhood with its waters. Dotted about the groves of fruit-trees and fields of maize and lucerne are numerous farmhouses, and about a mile from the town, in a pleasant garden whose centre is occupied by a reservoir thickly shaded by bending willows, is the camp of a *sotnia* of Cossacks. Rows of graceful young poplars mark the rills of irrigating water which render the country verdant, and which fill small ponds in plantations near the fort-like farm-houses.

These buildings are square enclosures of perhaps one hundred yards a side, whose smooth mud walls are about fifteen feet high, and have circular tower-like buttresses at the angles. Their interiors are divided into courts, which are devoted to the living rooms of the owner's family, or of the guests, or to straw and cattle yards. In the centre of the western wall is usually the entrance-gate, closed by heavy iron-bound folding-doors, over which the parapet wall is crenelated, to allow of a musketry fire being brought upon a possible attacking party. Such is the usual form in the Oxus countries of all homesteads which are situated without the walls of a town. In the Khivan oasis, where the population is chiefly employed upon agriculture, there are thousands of similar structures scattered along the banks of the irrigation canals fertilising the country. In type these buildings resemble the ancient habitations of the early Aryans as described in the 'Vendidad,' and since none are found beyond the neighbourhood of Cimbye, this place probably approximates to the northern limit of Aryan influence on the lands of Turan.

# CHAPTER XIV.

## THE LOWER AMÚ-DARYA—*continued.*

The desert east of Kuwán-jerma—Canals from the Kizzel-darya—Lake Dowkara—The Belitao hills—Yany-Su—Changes since 1859 in this channel and in Dowkara—Causes of these—The closing of the Loudon canal—Method resorted to by Oriental rulers for overawing feudatories since remote antiquity—Consequences of closing of Loudon—The Central Amú channels—Chertambye—Tullabye—Oguz—The story of Prince Bekovitch—Introduction of order and civilisation by Russia in Turkestan.

BELOW the head of the Kigailee canal the Kuwán-jerma arm of the Amú cuts, in one locality, almost at right angles, several high parallel ridges of sand; which appear to be the banks of old canals, carrying water in a general east and west direction. From the summits of these sand-ridges is seen to the east an extensive tract of rolling desert, in whose hollows grow lines of scrubby-looking trees; maintaining a hard struggle for existence in the arid soil that is moistened by a small quantity of water, which percolates into the low levels from the flood stream in Kuwán-jerma. Close to its banks there are large pools of water, supporting a luxuriant growth of rushes and other aquatic plants; but as the eye passes from these cool green masses to the glowing eastern horizon the country becomes more and more sterile, until the view is lost in the distant desert sands. It must have been somewhere across this neighbourhood that the Kizzel-darya, or continuation of the Syr, ran from Lake Kouktchatengis on the east towards the Uzboy channel of the Oxus, that flows down to the Caspian Sea. The Persian MS. in Sir H. Rawlinson's possession, which mentions this circumstance, whose curious and perhaps exaggerated result was described to be the desiccation of the basin of Aral, is corroborated by Uzbek annals, which allude generally to the course of the Kizzel-darya. The canals, whose traces have been above described, are probably those which formerly conveyed the waters of the Syr, for the fertilisation of the country near the head of Kuwán-jerma, and so absorbed that portion of the river's stream which formerly ran past the south-east corner of Lake Aral towards the Oxus.

From its head, near Khodjeili, the eastern arm of the Amú-darya flows in a rather meandering course, about one hundred miles in length, down to Lake Dowkara; a vast extent of lakes and swamps, having an area of nearly four hundred square miles, and traversed by various deep-water channels, of which the sole one navigable makes a long circuit of about sixty miles round its east and north shores. On the latter side Dowkara has a narrow bay, extending farther north into the land, and shut in by the cliffs of the Belitao range of hills, some two hundred feet in height, which extend easterly, in diminishing their elevation, along a line of fifty miles. At the south-western quarter of the mouth of the gulf enclosed by the Belitao cliffs, the Yany-Su leaves Dowkara by a curve to run north-easterly, along a distance of more than fifty miles, to its mouth in Lake Aral. The upper part of this channel passes through tolerably elevated and firm ground, covered with the usual scrubby jungle; but on its lower course, about the south-east corner of Aral, there is a plentiful growth of *saksaoul*. To the west of its mouth, as far as the Kazzakdarya, the country is little elevated above the surface of Aral, and is swampy, though it is studded with a few isolated hills of small height.

When ascending the Yany-Su in 1859 Admiral Boutakoff reached a point about twenty-five miles distant from Lake Aral, where the further progress of his steamer, the 'Peroffsky,' was arrested by a ridge of sandstone, which crossed the stream. The depths of water upon this rocky barrier were found to vary from 1½ to 2½ feet; that in the lower course of the Yany-Su having been generally five to eight feet. The 'Peroffsky' was consequently left at the ridge, and the ascent was continued in an open 12 h.p. steam launch for a distance of more than eight miles, along which the bottom was found to be stony, while sharp rocks rose here and there to nearly the surface of the water. Higher up the channel, close to the Belitao, the Yany-Su ran between stiff clay banks, rising from forty to sixty feet in height on each side of the channel, which had been excavated by the stream. This was found to flow from Lake Tampyne—a body of water separated from Dowkara by a narrow depressed spit of land, but communicating with it by a small connecting channel. At that time the two lakes were also joined temporarily, during the Amú floods, by the overflow taking place over the intervening low ground; but at the present day they form permanently a single body of water, and the name of Tampyne Ayage has disappeared in that of Dowkara, which also includes another lake on its east called Kungrad.[1] In 1874 the same vessel 'Peroffsky,' which has been mentioned above, steamed up the Yany-Su arm, and found no trace either of the sandstone ridge or of the other obstructions observed by Boutakoff in 1859. They have consequently been swept away by the action of the river in the interval, and the phenomena of change which have thus been in operation are good illustrations of the physical characteristics of the Amú-darya. As has been seen, the water-level of Dowkara is to-day higher than formerly, while that of Aral is lower; and hence the stream of this arm has now a greater fall. At the present day the Yany-Su has by no means reached its full development, though it carries down to its mouth in Aral perhaps three-eighths of the whole volume of water that passes Khodjeili.

Before proceeding to describe the central group of lakes and marshes of the lower Amú it will be convenient to glance at the circumstances to which their existence is chiefly due, and which have caused the enlargement of the Yany-Su arm of the Amú,

[1] No connection with the town of this name.

besides tending to change the whole flow of the river, from the western to the eastern outlet. In 1857 the Yomut Turkoman chief, Atta Murád Khan, the *aouls* of whose tribe are situated near the head of the now desiccated Gulf Abougir, declared war against his hereditary foe, the Khan of Khiva ; and the Uzbek Khan immediately retaliated by barring the head of the Loudon canal, which left the Amú-darya some few miles above Khodjeili, and so deprived the lands occupied by the Yomuts of the water necessary for their fertilisation. This operation of closing an irrigation canal for the purpose of overawing their feudatories has been commonly employed by the rulers of Khiva, and has, no doubt, contributed to many past changes in the flow of the waters of the Amú. It is the means which has been resorted to by Oriental rulers, since the remotest ages, for enforcing either a legitimate or illegitimate authority over subject populations, who are usually dependent upon an artificial supply of water for their very existence. As to this, the passage in Herodotus, which certainly refers to the country watered by the ancient Oxus, may be recalled, where the Persian king, yielding at last to the complaints and lamentations of the famine-threatened people, orders the sluice-gates of the river to be opened and lets the thirsty country drink until it has had enough. The historian adds: 'It has been told me, that the king never gives the order to open the sluices until the suppliants have paid him a large sum of money over and above the tribute.'

The considerable body of water which was formerly carried by Loudon having thus been thrown into the lower channels of the river, whose dimensions were quite unequal to the extra duty so cast upon them, the eastern arm began to enlarge itself. At the same time the western arm, in its course between Khodjeili and Kungrad, overflowed and detached new channels to get rid of its superfluous waters, which consequently formed the inundation now covering the central tract of the lower Amú. The largest of these channels is called Chertambye, whose head is situated at about twelve miles below Khodjeili. It runs almost due north for twenty-five miles before reaching the lake and swamp district near the Kashkanatao hills; and though its upper course (more particularly on the right bank) is tolerably dry and well-defined, its left bank and lower course are edged with marshes, which in the flood season extend to the west, and also pass insensibly into the group of rush-covered lakes below. As regards the increased quantity of water that was thrown into the lower Amú courses by the closing of Loudon, I was informed by a Kirghiz *moolla*, born and bred *in situ*, that the Chertambye channel had no existence a few years ago. This statement is confirmed by the map compiled in 1859 by Boutakoff, which shows the next westerly channel, named Karabailee, as the one situated most to the east in that year; though at the present day Karabailee is dry, while Chertambye carries about one fourth of the whole volume of flood water poured by the Amú into Lake Aral. The same *moolla*, when with me in a *caïque* immediately to the north of the Kashkanatao hills, stated that a dozen years previously he walked over cultivated fields which formed at the moment he spoke the bed of the water on whose surface we were. Some of the lines passed over in these localities by Boutakoff on foot during his early explorations have now to be traversed in boats; and this is easily accounted for, since in 1859, the second flood season after Loudon had been closed, sufficient time had not elapsed for the formation of the channels which to-day carry down water for the inundation of the country. Still Boutakoff himself noted in some places the commencement of the flooding, for while passing along the Karabailee channel he remarked the limpidity of the water, which allowed of submerged fields and irrigation channels being seen through it. Westerly of Chertambye there are at the present day only two channels, Tullabye and Oguz, which lie between it and Kungrad, though formerly there were others, besides Karabailee, whose desiccation has been mentioned. The change thus seen to have taken place is probably due to the enlargement of Chertambye and Kuwán-jerma, and indicates the general tendency of the Amú to pass to Lake Aral by its eastern outlets.

On the north of the central lake district is a hill called Burlytao, as well as some low rising ground, situated among the marshes, and called Bukhlitao; though the latter could only have pretension to being a '*tao*,' i.e. hill, if the inundation were drained off. On the west, the Tumalaktao hill has already been described, and from this the whole country eastwards is converted in the flood season into marshes connected with the central lakes.

During the summer of 1874 I passed from the upper Ulkun, near Tumalaktao, into the Oguz channel, the most westerly of those feeding the lake district. The passage was made by dragging the *caïque* for nearly a whole day through a forest of rushes, which grew out of three or four feet of water, whose bed was a mass of tangled weeds. Eventually an open lake upon the course of Oguz was reached, and the banks of this channel were found to be rather drier, and occupied by an *aoul* of Karakalpaks. Oguz is certainly a very old channel of the river, and it was found to be very deep in proportion to its breadth—a circumstance probably due to an increased volume of water having been projected into it during floods of late years, and having acted in excavating its bed rather than its banks. Tradition says that in the year 1716, upon the banks of Oguz was enacted the tragedy in which the Prince Bekovitch played the chief part, and of which the circumstances are briefly as follows:

Bekovitch was a Circassian prince, and an officer of the Imperial Guard, who had been converted from Mohammedanism. He was sent by Peter the Great to explore the country between the Caspian and the Aral, and to enter into commercial relations with the Khanates of Central Asia, and was placed in command of a force of some three thousand men, the greater number of whom he led into Khiva, after having examined the old Uzboy channel of the Oxus. The Khan at first opposed the entry of the Russians into his territories, but having been won over by presents, and hoping for further ones, he allowed Bekovitch and the men under his orders to establish themselves in Khiva, preparatory to a future advance into Central Asia.

The cupidity of the Khan having been once aroused quarrels soon broke out, in consequence of Bekovitch declining to make more presents, and the Uzbek chief accordingly formed the design of avenging his disappointment, and at the same time of getting rid of his unwelcome guests. Under the pretence of the difficulty of furnishing supplies to so large a number of people, collected in a single locality, Bekovitch was induced to divide his force into many small detachments, which were scattered over the face of the country. In these isolated positions the several parties were fallen on, one after another, and destroyed in detail by the Uzbeks.

Prince Bekovitch himself, being the object of the Khan's personal hatred, was flayed alive before being murdered; while his two Mohammedan brothers, who were offered their lives on certain conditions, elected to die with Bekovitch rather than dishonour themselves. Of the whole number of the Russian expedition it is said one man only, who was in some way suspected of having played the traitor, escaped to tell the story, for whose completion it should be added that the Khivan Khan offered his excuses to Peter the Great subsequently, in affirming that the massacre was rendered necessary by Bekovitch's bad faith in keeping back the presents sent by the Czar!

Had not Peter the Great been fully engaged about this time in extending his empire on the shores of the Baltic and in his wars with the Swedes, the Uzbeks of Khiva would probably have paid dearly for the murder of Bekovitch and his men. As it happened, it was not till 1724 that the Czar again turned his thoughts to the Caspian regions and took Derbend, his last military exploit. He died in the following year, leaving as a legacy to his successors that policy of aggrandisement in the East which the present century has seen so tenaciously followed and successfully executed. Though Russia's position in the Central Asian

Khanates may not yet be assured, it is certain that without her leave no dog may bark in the bazaars of Khiva, Bokhara, and Kokand. And if a strong government which preserves social order and has put down brigandage, slavery, and man-stealing is worthy of sympathy, it is impossible not to feel that in undertaking the thankless and costly task of introducing civilisation into Turkestan, Russia is fully entitled to the good wishes and gratitude of every Christian nation.

# CHAPTER XV.

## THE CENTRAL AMÚ SWAMPS.

The lake district and the northern outlets of the Araxe of Herodotus —The central depression of the lower Amú—The whirlpools—Lakes and connecting channels—The Kashkanatao hills—Former water-level of Aral—The Karakalpaks—Their poverty—Poor pastures—Agriculture—Locusts—Fisheries—Musquitoes—Karakalpak traits—Russian policy with Turkestan races—The leper-woman—Kirghiz on lower Amú—Ethnic affinities of Russia with Asia.

As the lakes and swamps formed by the central channels, by which the Amú-darya carries a portion of its stream to Lake Aral, are passed through, it is impossible not to be struck with the resemblance of these watery precincts to the description of the northern outlets of the Araxe of Herodotus. The Father of History himself might well have passed in a *caïque* between the walls of tall rushes which border these channels, or have landed on some of the many half-floating islands on which the Karakalpak fishermen establish a temporary resting-place. These people, who are hard put to it for fuel of any kind save the stems of the rushes they store up for winter use, might well be the descendants of that ancient race who dwelt in the swamps formed by the forty mouths save one of the Massagetic Araxe, and who fed upon raw fish; and though the Karakalpaks are not now-a-days clothed in sealskins, as the people were whom Herodotus describes, they might have been so little more than a quarter of a century ago, when these animals were not yet extinct in Lake Aral.

The great region of lakes and marshes is comprised between the Ulkun-darya on the west and the Kashkanatao hills on the east; but in the flood season an extensive area of drowned land is added on the north, where the country on the right bank of the Amú-darya, for nearly the whole distance between Khodjeili and Kungrad, is entirely inundated. Though the fall of the ground in the direction of Aral is considerable, this water-laden tract presents the appearance of a large sea, whose surface is at rest, except in the different open channels which traverse the thick aquatic growths of giant *arundo* and *tipha.*

The more permanent portion of the lake and swamp district is included in a caldron-like depression, which, according to Russian information of 1641, seems at that date to have been entirely dry. On the east, as well as on the west and south, though in a lesser degree, the caldron is skirted with elevations, by whose upheaval it was possibly caused, and which would have been islands in a Lake Aral whose surface was fifty or sixty feet higher. A little to the north of the swamps is a curious phenomenon, which may also be a result of that system of subterranean disturbance whose traces, geologists say, exist along a line drawn from the east of Asia, through Lake Aral and the Caspian, to the countries bordering the Black Sea. This phenomenon occurs at the junction of some channels draining the central depression into the Ulkun-darya; and here are situated, side by side in the stiff clay subsoil, two deep cavities shaped like inverted cones, in which the stream forms whirlpools. Quantities of fish of large size sport in these foaming waters, over whose troubled surface flights of gulls and other aquatic birds hover and circle in search of their prey. There is no reason for supposing that these deep inverted cones have been excavated by the action of running water; and in the absence of any guiding indications it seems only possible to class them as craters of extinct mud-volcanoes, such as those of Lussbeyla, near the mouths of the Indus, or as analogous phenomena to the circular hollows left by the Calabrian earthquakes of 1783–86, which are described in Lyell's *Principles of Geology.*

To return, however, to the central lakes. These are sheets of water varying in area from a few thousands of yards to ten or fifteen miles, and their *maximum* depth may be as much as thirty feet, though the average is about seven. Their beds are for the most part a mass of weeds, composed chiefly of the delicate *miriophyllum*, from which lotus plants of different-coloured flowers rise to the surface of the water. Little current is observable except at the actual outlets or inlets of the lakes and in the channels, where sometimes water flowing with a high velocity carries along a floating island of rushes, which has been torn up from its place by an extraordinary flood. The breadths and depths of these connecting channels are of all dimensions, the *maximum* of each being, perhaps, one hundred yards and twenty-five feet respectively. Both lakes and channels are walled in by dense growths of gigantic rushes, attaining almost to the dignity of trees, and adding immensely to the sense of solitude and mystery which pervades these uncommon scenes.

On the east the lakes are bounded by the Kashkanatao hills, whose summit is occupied by a plateau at an elevation of about three hundred feet, which extends in a north-east and south-west direction along a distance of twelve miles, with a breadth of about two miles. The hills are of indurated clay of a buff colour, full of pieces of selenite; and their rounded flanks are covered with sand and small pieces of ferruginous sandstone. Their surface is sprinkled over with a few thorny shrubs and prickly plants, such as *lycium*, *haliostachis*, &c., whose dry and sapless foliage is suitable only for camels. At the south-western extremity of the plateau (and therefore sheltered from local north-east gales) are several detached elevations in the form of rounded hillocks, smoothly worn by water, and encircled with horizontal water-marks and narrow beach-like terraces. These beaches, as well as the surfaces of the promontories which connect some of the hillocks with the main body of Kashkanatao, are covered with thin sedimentary layers of sand, that are hardened with iron and frequently marked with distinct ripple-marks. Such appearances have evidently been caused by the rising and falling surface of Lake Aral, when that body of water had its level about sixty feet above that of to-day.

The people who inhabit the central lake district of the lower Amú-darya are the Karakalpaks—a tribe of Mongol-Turks, whose name is derived from a black felt hat they are supposed to wear, but which in practice they seldom do wear. At the breaking-up of the Golden Horde of Kipchak this tribe occupied the pasturages of the lower Syr-darya, but as they kept aloof and joined neither the Kirghiz hordes nor the Uzbek confederacy they soon found themselves in the unenviable position that is usually occupied by a minority in this world. While the Uzbeks had greater things to do, and marched to the conquest of the Central Asian Khanates, the Kirghiz found themselves powerful enough to encroach upon the Karakalpak lands, and eventually compelled them to move on to the lower courses of the Amú-darya. The numbers of this tribe are at the present day 50,000 souls, at a very outside calculation, of which the great majority live on the lower Amú though some few are established on the right bank of the river, higher up, near Khodjeili.

During the past eighteen years, since the closing

of the Loudon canal and the resulting yearly inundation of their pasturages, the material position of the Karakalpaks has very much altered for the worse. From their first appearance on the Amú, they were always on bad terms with the Uzbeks; and having been ill-advised enough to ally themselves with the Turkoman Chief Atta Murád Khan against the Khivan Khan, they soon paid the penalty of such a foolish proceeding. The Uzbeks killed two birds with one stone, by barring the Loudon canal, and so at the same time deprived the lands of the Yomuts of the water necessary for their cultivation and flooded the pasturages of the Karakalpaks. A large part of the latter country having thus been changed into swamps, their cattle, the sole wealth of a pastoral race, have greatly diminished in number, since even much of the remaining land is in a half-drowned state and makes but very poor grazing ground. As to this, I remember seeing one day a herd of brood mares with their foals highly excited by the unwonted appearance of a boat-load of Europeans, and floundering belly-deep in the muddy pastures which lined the banks of the stream we were passing along at the time. At the present day the Karakalpaks are among the poorest of the Turkestan races, and the occupation of the Amú district by Russia can scarcely fail to benefit them very materially.

Though the great majority of these people are still pastoral shepherds, some few are sedentary in Kungrad and Cimbye, while others devote themselves to agriculture of a sufficiently rude description. As may be gathered, there is no great extent of cultivation, but it is difficult to see how such as there is can escape the ravages of the enormous clouds of locusts, like those which daily darkened the sky during July 1874. One of such clouds was estimated to measure fifteen miles in length, by two miles in breadth, and to have a depth of half a mile; and, allowing two locusts only for every cubic yard, there would have been more than 38,000,000,000 of these destructive insects in this flight. However, this is a very moderate estimate of their probable numbers, which might very well have been twenty times as great. The vast, bare steppes and deserts of the Aralo-Caspian region, that possess little moisture and offer their surfaces to the full heat of the sun—both of which conditions are necessary for the hatching of the eggs deposited in the soil—afford the best situations for the propagation of locusts, whose increase is also encouraged by the absence of forests, in which insectivorous birds could find a shelter. It might be inferred that the Khivan oasis must be exposed to great danger from this plague; but in practice the large number of small birds in the planted groves of the Khanate, seem to afford a sufficient safeguard against the ravages of locusts.

A great number of Karakalpaks are fishermen, who take, in fixed nets, quantities of a large, coarse sturgeon, with which the waters of the Amú abound, and which, dried and salted, form the staple of a very brisk trade carried on by the boats of the Amú and its branches, for distribution among the nomads of the Khwarezmian deserts and the sedentary populations of Central Asia. These fishermen pass most of their time on the semi-floating islands in the central lakes, where a footing is formed only by the tangled mass of interlaced roots of the *arundo* and other aquatic plants. Poor and ill-clothed as these people are, each little gang of fishermen has a canopy or tent of cotton cloth, within which to pass their nights; for without such shelter, sleep, and perhaps, indeed, life, would be impossible, so innumerable are the mosquitoes and so painful are the bites of these insects in this locality. Mosquitoes are serious evils in many other parts of the world, and stories have been told of seamen driven to jump overboard, and so to commit suicide on this account, in the Rangoon river; but it may be doubted whether any more exquisite torture can be suffered than that inflicted by the mosquitoes of the lower Amú. What it really amounts to may, perhaps, be indicated from the fact that, all hardy and insensible to bodily evils as the Russian soldier and sailor is, even *he* must be provided with mosquito-curtains in this part of the world. Like the insects of the pretertiary epoch, which M. Edgar Quinet described, every form of instrument of torture seems to be affected to the use of the Amú mosquitoes—teeth, drills, augers, saws, tweezers, barbs, pincers, files, and hooks—each lends an acute thrill of novelty to the pain of every successive bite, and prevents the sufferer's sensations becoming deadened by a mere sameness of suffering.

The Karakalpaks are a rough but cheery set of savages, and, though not churlish, are probably too poor to be hospitable. Their dealings with their new Russian masters seem to be carried on by the medium of a few more civilized Kirghiz *employés*, drawn from the Syr-darya districts; and, though the Karakalpak is despised by other peoples of Turkestan, he wears an air of independence and manliness when addressing a European. This trait among them struck me, indeed, very forcibly one day, when a young Karakalpak boy, who had a hawk perched on his naked shoulder, came up to inspect my proceedings as I sat sketching the *aoul* where he lived. Wishing to see what he was made of, I offered him a bright new twenty-copek piece, which he refused with such a glance of defiance that I was puzzled to decide whether the bird or the boy had the wildest or more untameable eye.

In her first dealings with the subject races of Turkestan, Russia has, not unwisely, restricted her policy to the preservation of social order, and to educational measures, which mostly aim at the dissemination of a knowledge of her language. Nothing but obedience to the laws is forced or obligatory, and old customs and prejudices remain for the most part uninterfered with; though the practical effect of such conditions, limited as they are, must be encouragement to the nomads to become sedentary and to cultivate lands more nearly under the protection of military posts. In this way, though her progress may be slow, her dominion in the future is assured, while the inhabitants of the independent

Khanates, who are subject to the arbitrary laws and despotic whims of their rulers, are surely being prepared to welcome the Russian rule.

The inconvenient side to this policy from a higher point of view is that many barbarous habits and superstitions, which are repugnant to more civilised and humane ideas, are left uninterfered with. One day, on the lower Amú, I came upon a Karakalpak woman in the jungle, who crouched upon the ground, implored food with a haggard face, in stretching out a pair of bony arms. The circumstance was unusual enough to attract my curiosity, and on enquiry I found she had been expelled from her *aoul* to die of hunger, as her body showed the marks of Arabian leprosy—a disease accounted in Turkestan to be a mark of Divine displeasure. I sent the poor wretch some food, but the Tartar who carried it showed signs of abject terror as he approached the spot where the woman was. Taking courage, he advanced, repeating an abundance of prayers; but, giving way at least fifty yards from the woman, he took to his heels, in an agony of superstitious fear. The boatmen of our party informed the Karakalpaks how the woman was being kept alive, and a day or two after this she was removed to a more secluded spot in the jungle, where she was certain to die of starvation, without the interference of meddlesome European strangers.

In addition to the Karakalpaks, some Kirghiz are found to the west of the Yany-Su arm of the Amú, which, however, may be said to be the general line of demarcation between the two peoples. Such few Kirghiz as are met without their more regular limits, are probably descendants of those who have been for some time subjects of Khiva, and who had, perhaps, possession of the ground previous to the more recent arrival of the Karakalpaks from the eastern coast of Lake Aral. The *moolla* mentioned on a previous page as acting as guide to our party on the Amú was one of these Kirghiz; and, besides farming land, he had a fishery in which he employed Karakalpaks. In addition to such mundane occupations he was the religious guide of the greater number of the local population, and, no doubt, received a snug revenue when upon his periodic tours to celebrate marriages and other rites. Being personally acquainted with every man, woman, and child, and every square yard of the country, no better *cicerone* could have been had. He was a jolly, easy-going ecclesiastic of the Friar Tuck variety, who knew all the best melon-grounds and orchards, and where the fattest sheep were to be had, and was a welcome guest at every *kibitka* in which he might appear. He was very regular with his prayers five times a day, and would, certainly, without that resource for killing time, have been bored to death; cut off as he was, during the eight or ten weeks he was with us, from the usual round of Kirghiz and Karakalpak social life. It was only, indeed, in the evening, when the *koumiss*[1] skin circulated round the circle seated about the roaring camp-fire, and when his servant (an accomplished local musician) made night terrible with songs relating to the merits of the 'noble Bokhara,' that the genial *moolla* made up for the *ennui* and the worry he had suffered during the day. At such a moment the dancing flame-light lit up features which passed insensibly from the Uzbek and the Kirghiz to the Tartar and the Cossack, and found a reflection in more Europeanized Russian eyes, which sparkled to the surroundings of the scene. It is then that a conviction is felt of the error made in

[1] Fermented mare's milk.

talking about Russian aggression in Asia, and that the appropriateness of the feeling can be recognised which urges the old national party to a return to the historic homes of a great proportion of the population.

# THE LOWER ARMS OF THE AMÚ-DARYA

To face page 198

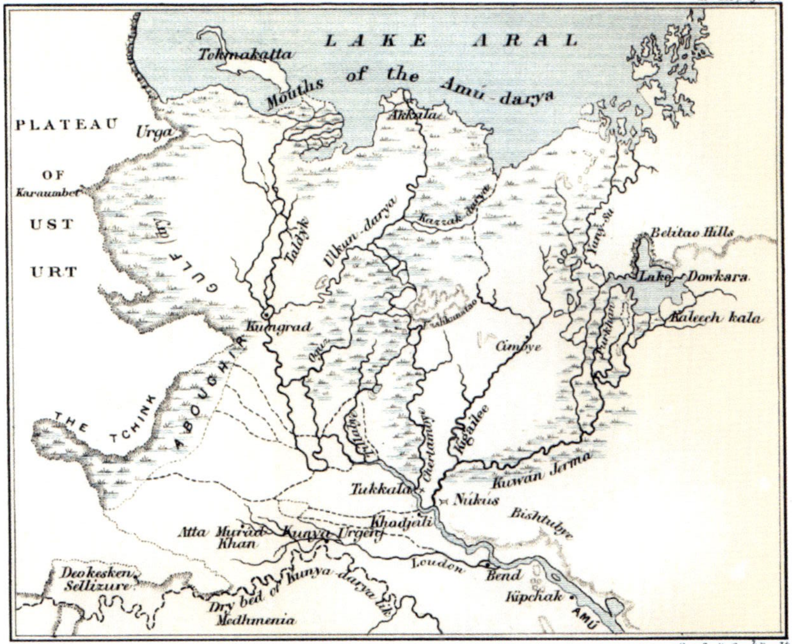


Scale of Miles
10 5 0 10 20 30

# CHAPTER XVI.

## CHARACTERISTICS OF THE AMÚ.

The first sight of the Amú-darya—Its banks from Khodjeili to Toyou-boyin—Breadths of the river—The deep channel—Current of river—Process of change going on in the channel—Illustration of this from M. Vámbéry's description of the Amú in 1863—The aspects of the channel at high flood and subsequent to a fall in level—Phenomena of flow of the Oxus—Tendency to re-establish conditions disturbed by Man—Inability of population to control the flow of the Amú—Future change.

On arriving at the point above where the Amú-darya divides into its lower branches the view of the great yellowish-brown stream, flowing on with scarcely a sign of the human movement which is usually characteristic of great rivers, seems to place the faculties of the spectator at once in sympathy with the attributes of the ancient Oxus. That mystery, too, which has in all modern times attached to the desert-surrounded oasis of Khiva, seems embodied behind the thick screen of elms and poplars, planes and mulberries, which follows at no great distance the windings of the river, upon its left or western bank. Upon the opposite bank, except where irrigation is also practised, over a limited area, some distance further up the stream, the sterility of the Kizzelkoom desert is shut out by the bare clay slopes of the Bishtubye ridge. This continuous line of low hills commences opposite to and a little above Khodjeili, and, continuing some sixty miles south, runs into the lower spurs of the sombre-looking mass of the Shaikhjaili mountains. Fringing both banks, nearly the whole way so far, is a vigorous growth of jidda (*eleagnus*) jungle, and along one hundred miles still farther south this is replaced by cultivation, which on the Khivan bank is brought to a high degree of perfection. In these hundred miles of the river's course the channel is much widened, and is occupied by numerous and large alluvial islands, which are covered with scattered young willows, hidden in the summer, by thick crops of high golden-eared waving grass (*lusiagrostis splendens*). Along the Amú-darya, in this part, are detached from the left bank many large irrigation canals, which water the Khivan oasis, and some few smaller ones also fertilise the Russian territory on the right bank. Among such channels are also found traces of older canals, which formerly carried the waters of the river; and more particularly on the left bank are at least two, of dimensions sufficiently large to have been main courses of the Amú itself.

Above the broad portion of the river the Kizzel-koom desert gradually closes in, on the cultivation of the east bank, while on the west bank that of Khiva is replaced by a high, bare ridge of clay, passing almost into an argillaceous schist, which the waters of the stream have cut away along some miles into a steep, though not very lofty, precipice. At the upper end of this ridge, about one hundred and ninety miles from Khodjeili, is Toyouboyin, 'the Neck of the Camel,' where the Amú, flowing from the east, turns south-west and rushes through a narrow passage, which its waters have cut through a bed of compact limestone. This dyke of rock is full of small shells, and is of the age of the chalk, and the channel in which the river flows through it has a breadth of about eleven hundred feet. The rock has been tilted to a height of about twenty feet, directly across the direction of the river, and its strike is thus about north-west and south-east, while the angle of its dip is, perhaps, 25°.

From Toyouboyin downwards, for a distance of

more than twenty-five miles, to Tunuklu, the breadth of the Amú increases to about two thousand feet, after which, in the portion where water is diverted for irrigation and where the larger islands occur, the river has, perhaps, as much as ten thousand feet between its banks. Below this, again, the breadth decreases, progressively yet irregularly, to fifteen hundred feet at Khodjeili and to twelve hundred lower down towards Kungrad.

The deep-water channel has a depth during the flood season of from fifteen to thirty feet, according to the locality where it is measured, and is generally found close to the right bank of the stream. This is due to the fact that the Amú, flowing to the north in the northern hemisphere, submits to the action of the great astronomical laws, and its current, acting in conjunction with the axial rotation of the earth, erodes its right bank. So generally the deep-water channel is found close to this bank, except where the natural flow of the river is interfered with by the action of the streams diverted into the irrigation canals. In the portion where the breadth of the Amú is great the deep-water channel not unfrequently, in consequence, abandons the right bank and flows in a winding course, between the alluvial islands;

thus adding considerable difficulties to the steam navigation of the river.

The current of the Amú in the flood season is variable but high, and amounts in some instances to as much as six miles an hour. The variations are due to the different slopes of the several portions of the bed; to the diversion at various points of large streams of water into the irrigation canals, and to the varying breadths of the channel. In a high flood the broader reaches become reservoirs in which water is stored up, while the narrow reaches connecting them carry streams running with a high velocity, which act to enlarge these channels. Such enlargement takes place, however, at different points, according to the quality of the material being acted upon; and the velocity of the river is, therefore, influenced by the circumstance of its passage through the limestone rock of Toyouboyin, the metamorphic schists of Shaikhjaili, or the several clay and sandstone formations of different degrees of hardness which occur along the course of the Amú. These irregular conditions of flow show that even at the present day, more than two centuries since the river has been running in the one hundred and twenty miles of channel above Khodjeili, the enlargement of its channel is not yet sufficiently completed and its ultimate *régime* has not yet been attained. The actual work in process of being executed in the channel of the Amú may be judged of from what has been already described, as having recently taken place in the Yany-Su arm, and receives further illustration when its present conditions are compared with those that were observed and recorded by M. Vámbéry.

These are more particularly referred to in two passages which at first sight would seem to be very inaccurate, but which are of the highest value, in showing the phenomena that characterise the flow of the river. In one passage M. Vámbéry describes a waterfall to exist near Khodjeili, where the stream rushed down from a height of three feet with the swiftness of an arrow, and so formed an obstruction to the navigation of the Amú. From what has been said regarding the river it is easy to understand that in 1863, the date here in question, the Amú must have been working actively in enlarging its waterway in order to pass off that additional stream which had been added to it by the closing of the Loudon canal in 1857. The aspects referred to can, therefore, be partially accounted for, though it is impossible to state precisely what this obstruction was, which in another place M. Vámbéry calls a dam. However, it is certain that for some years subsequent to 1863, a portion of the Amú still flowed away to the north-east, by the Karauzak channel, whose head was exactly opposite to Khodjeili, and a little above that of the present eastern arm of the river. Karauzak was closed by the Khivan Khan, to punish the nomadic tribes living along its course; and the dam mentioned by Vámbéry may, very probably, have been a spur of clay and brushwood, which was erected at the head of the channel to deflect the stream from entering its mouth. In such a manner a deposition of earthy matters would have been induced about this spot, and the canal would have been closed as required. At the present day no waterfall exists and no trace whatever of such a dam is to be seen. But there would have been no difficulty in its construction, and a spur or groyne of the kind would have differed in dimensions only from hundreds of similar ones seen along the banks of the Amú, and in the still water behind which the fishermen fix their nets.

A second passage contains M. Vámbéry's description of the river, at the head of the reach skirting the Shaikhjaili range of mountains, where the waters were said to roar, as if the Oxus, that unruly son of the desert, were angry at being so imprisoned between the rocks. Below this point the stream was described as having a narrow channel, dangerous to navigation from the force and rapidity of its current. If such a state of things exists no longer, the passage is valuable, as showing the change which has taken place at this point; for, though the place in question is situated in one of the narrow reaches of the Amú, since 1863 the channel has been sufficiently enlarged to remove all danger, if not, indeed, all inconvenience to the navigation, and the roar of the waters which Vámbéry spoke of is now hushed for ever.

I had the opportunity of seeing the Amú very thoroughly at the epoch of the highest flood of 1874, as well as five weeks subsequently, when the level of the water had fallen five feet. At this latter date, in lieu of a single sheet of water, the broader parts were encumbered with numerous exposed sandbanks, between which meandered the several channels making up the whole stream of the Amú. In the narrow connecting reaches the river still flowed in an unbroken stream, but with diminished breadth and a moderate current in place of the large volume of water whose velocity at the flood-level was so great that the steamer 'Peroffsky' could sometimes do little more than hold her own against it in her ascent. Such was more especially the case in the reach immediately below the head of the Loudon canal, whose barring in 1857 threw a very large additional volume into the lower courses of the Amú. The entire inability of this part of the channel to carry the flood volume was very manifest, since time has not yet elapsed to allow of its sufficient enlargement to pass the present stream with ease down to an outlet.

In the conditions of flow of the Amú-darya we have a striking, if not, indeed, a unique, example of natural forces unceasingly seeking to restore an equilibrium which has been subverted by the interference of Man. In the struggle for existence the populations of the Khwarezmian plain have, at various epochs, commencing, probably, with the remotest antiquity, succeeded in fertilising large areas, by the water of canals led from the Oxus, along the higher ridges of the country. But the full development of science was rendered impossible by waves of barbarian invasion, which, century after century, swept down from the far north-east of Asia. The civilisations current in these harassed regions, have never included the knowledge necessary to ensure the permanency of the system of irrigation which had been created; and this has consequently been always under the domination of hydraulic phenomena, incessantly working to re-establish the flow of the Oxus in a manner more conformable to the laws that Nature had ordained. From its original mouth in the Caspian, the Oxus on this account, after several oscillations, now flows to an outlet which is situated upon a much higher level; by a channel whose *régime* in its lower courses is still far from having reached its full development. Until this result has been achieved no further radical change need be apprehended; but immediately the prevalent excavating action of the flood-stream shall be replaced by a slower velocity, and by consequent earthy deposition in the lower course, a new outlet will be sought for by the Amú, whose waters, will in that case infallibly find their way once more towards the Caspian. The process by which this change of outlet is effected, will be explained in detail in the succeeding chapter.

# CHAPTER XVII.

## THE AMÚ IN KHIVA.

The Amú-darya course from Lake Victoria—How it waters the Khwarezmian desert—Its volume—Khiva a gift of the Amú—Water used in irrigation of the oasis—Slope from river to the west—The canals of Khiva—Their action in causing change in the flow of the Amú—Cultivation of the oasis—Revenue of Khiva from land and other sources—Population—Facilities for settling the wandering Turkomans of the Khwarezmian desert.

THE Amú-darya, the Jihún of the Orientals, and Oxus of ancient historians, has the highest of its sources in Lake Victoria, which is situated upon the great Central Asian plateau, called Pamir. Running westwards, with a little southing, by a precipitous course, along a distance of about two hundred miles, through the mountain principality of Wakhan, it turns north to describe a semi-circular curve, about two hundred and fifty miles in length, round Badakhshan. In this part of its flow the river is joined by several large affluents, that descend from Pamir and from the southern slopes of the Karatao mountains, which bound the Zarafshan valley on the south. It then continues to the west, receiving drainage from both sides, along a distance of three hundred miles, after which it turns more and more to the north and enters Lake Aral, at about fifteen hundred miles from its sources.

It is a common thing to talk about a country being watered by a river, yet in the majority of cases the expression is not very accurate, for rivers chiefly serve to carry down drainage to the sea. When, however, we come to the almost rainless tracts lying to the east of the Caspian, which are rendered habitable and fertile solely by the employ of the waters of the Amú-darya, the force of such a phrase can be fully appreciated. This country has in all history been described as a desert, but there are, perhaps, grounds for believing that in a yet more remote antiquity it was the home of early Aryan peoples, and in as flourishing a condition as that of the oasis of Khiva at the present day. The explanation which will be submitted of how changes in the main course of the river crossing this desert are brought about by the diversion of water for the fertilisation of the country upon its banks, will go far to show that the first change in the flow of the Oxus was due to similar phenomena that were called into action during a prehistoric civilisation. And, as will be seen, the relics of this are actually to be found in that very sterility which has characterised these localities during the past twenty-three centuries.

But such considerations must be reserved for the moment to allow of the description of the hydraulic character of the Amú-darya, which in the lower three hundred miles of its course is employed in fertilising the oasis of Khiva, by means of a number of large canals diverting the waters over the surface of the land. The cultivating season begins with the early spring of the year, when the snows and glaciers about the higher sources of the river and its affluents commence to melt and to swell the volume of water flowing in the channel. This flood increases in a variable manner to a maximum, which in 1874 was attained at Khodjeili on August 3; and it may therefore be inferred that the epoch of greatest summer heat on the Pamir plateau, occurred some ten days previously. On the above day of extreme flood the discharge of the Amú amounted approximately to 143,000 cubic feet per second, a quantity which is almost twenty-one times greater than the flood volume of the river Thames at Staines. The lowest level subsequently reached by the Amú was on March 22, 1875, when the volume was approximately 35,000 cubic feet per second. As far as could be judged, the year 1874 was one of average discharge.

While Egypt owes its fertility to the waters of the Nile, the very existence of the Khanate of Khiva depends on the floods of the Amú-darya. It is, however, to be remarked that the fertilising principle which is contained in the waters of the Nile, and which is due to vegetable matter carried down from its upper courses, is wanting to those of the Amú. The river, in fact, brings down nothing but finely-triturated sand, the *débris* of that extensive destruction which Wood has described as caused by frost on the steep flanks of the mountains of the upper Oxus. It may consequently be inferred that the fertility of the Khivan oasis is due to the stimulation by water of the chemical properties of the clay which chiefly forms the great Khwarezmian plain.

Of the whole volume of water passing down the Amú in the cultivating season, about one-half, roughly, is diverted by the several irrigation canals watering the territory of Khiva, leaving the other half to flow down, into Lake Aral, from whose surface it is lost in evaporation. Excepting the few canals on the right bank, nearly all are situated on the left bank; from which the country slopes in a westerly direction down to the furrow that follows the foot of the Ust Urt plateau and formerly received the southerly overflow of Lake Aral by the head of the Abougir Gulf.

Since it is impossible to foresee whether the annual flood of the Amú will be a high or a low or merely an ordinary one, and since a certain minimum quantity of water is required for the cultivation of the crops, without which the Khivan population, isolated by surrounding deserts, would infallibly starve, the beds of the irrigation canals at their heads must be adjusted on such a level as will ensure the necessary quantity of water entering them, even in a year of lowest flood. It consequently results that in all other years the volume of water diverted by the canals for the irrigation of the oasis is larger than what is actually required, and this surplus, which would be as great an evil as a deficiency of water, must be got rid of as speedily as possible, in order to save the growing crops from destruction.

From what has been said of the general slope of the country towards the west it will be seen that there is every facility for the cultivated tract being inundated from the Amú, and escape channels are therefore dug to carry off surplus water into all adjacent low lands; and in this way many lakes and swamps are formed over the surface of the country. The canals are also lined with protective embankments, and the whole existence of the agricultural population, who live in homesteads scattered along the various watercourses, is devoted to the repair and maintenance of the works which guarantee the safety of the irrigated and cultivated tract from inundation. The entire mode of life of the people may, indeed, be said to be influenced by the phenomena of the Amú-darya, for at the end of the cultivating season and at the approach of winter, after the cereal crops have been reaped and garnered, and when the volume of water flowing in the river is greatly diminished, dams are built across the heads of the canals to prevent the further entry of water. After having in this manner been run dry, the earthy matter, amounting to a depth of more than two feet, which has been deposited during the year by the streams flowing in the canals, is excavated from their beds by a *levée en masse* of the population. By these precautionary measures the dimensions of the canals of irrigation are restored, and the entry is ensured of that minimum volume of Amú water which is required at the advent of the next year's flood, for the cultivation of the oasis.

The enormous mass of earthy matter which the Amú carries down in its descent from the high plateaux of Central Asia would, in a state of natural flow, have been conveyed to a permanent outlet; but since the transporting power possessed by the river is diminished by its division into canals of irrigation, a large portion of the earthy burden is deposited in the several canals, which divert one-half of the waters, as well as in the main channel, which carries the other half to an outlet. It has been said that the canals are cleaned out yearly, and are thus freed from these deposits; but those in the lower portion of the river are, year after year, continually added to, and as this increase of earthy matter takes place, and the main channel is gradually choked, a tendency must also be gradually and continually growing for more and more water to enter the irrigation canals. The cultivators have to combat the results of this evil, by which their fields would be inundated, by extending the courses of the canals, year after year, and getting rid of the water by a larger number of suitable escapes; and since opportunities for effecting this would occur in greater numbers along some one particular canal, this one would offer greater facilities for the entry of water from the Amú, and would become the chief channel by which the increased flow from the river would eventually declare itself. Thus this channel would be continually enlarged, until its dimensions first gradually reached, and then surpassed, those of the main course of the river, and the day arrived when the whole stream would inevitably change its flow into the enlarged canal, which would then become the new lower course of the Amú. The country watered by the canals which left the river in its old course below the enlarged one would consequently be deprived of the water necessary for its fertilisation; and the population so affected would change their habitations to some suitable ground upon the new course. In the phenomena described is to be found a sufficient explanation of the changes which the flow of the Oxus has undergone since time immemorial; and its application to the other rivers of Western Turkestan affords a key to the disappearance of much water which formerly entered the basin of Lake Aral, and caused its junction by overflow with the Caspian.

The clay soil of Khiva, more particularly in the southern parts, produces under irrigation, cotton of excellent quality, which Vámbéry's observations, confirmed by local report, show to be unequalled by any grown elsewhere in Central Asia.[1] On the other hand, the silk of Khiva is not so good as that either of Bokhara or of Kokand. The other crops grown on the Amú are wheat, maize, rice, lucerne, sesamum, hemp, and the madder plant, besides tobacco and the poppy, on the growth of which narcotics exceptionally high rates of land-tax are levied. Tree culture is largely practised, and the irrigated tract is covered with groves of mulberry, and with fruit orchards, as well as with poplars, elms, willows, planes, and other useful trees, which are planted along the banks of the canals and the network of minor water-courses.

The area of the sandy desert, which is more or

[1] The profit on an acre of Turkestan cotton is stated to be 5*l*.

less reclaimed by the waters of the Amú, is about one and a half million of acres, for the cultivation of which probably one-half of the volume of water that is now diverted from the river in the flood season would suffice. The Khivan landlord receives one-fifth of the produce as rent, but where irrigating water has to be lifted, the proportion is reduced to one-seventh. Of the whole oasis, perhaps one quarter pays a land-tax of about a shilling an acre to the Khan, the remainder being the property of religious or charitable corporations, or otherwise free from charge. In addition to this source of income an *ad valorem* duty of 2½ per cent. is levied on all imports, and the tributary Turkomans are supposed to pay eightpence yearly for each camel and bullock, and fourpence for each sheep they possess. The receipts from this tax must have been precarious previous to 1874, when the Khan's authority was scarcely recognised by the Turkomans; but the presence of the Russian battalions in the Amú-darya territory now probably acts as a pressure on these tribes, and allows of the indemnity being collected which was levied on Khiva at the close of the campaign of 1873.

The country's wealth has never recovered the

blows it received at the hands of the plundering hordes of Djengiz and of Timúr. At present, the usual revenues of the Khan are stated to amount to about 30,000*l.* a year, a sum far short of that in the old days, when, escorted by five thousand horse, he went on his annual hawking tour and collected tribute from the south-east coast of the Caspian to the courses of the lower Syr, and from the environs of Balkh to the Emba steppe, on the north of the Ust Urt plateau. Besides such more or less legitimate sources of revenue to the Khan of Khiva, belonged by right a share of the booty taken in the yearly *alamáns*, or plundering and slaving expeditions, formerly made upon Persia and the Kirghiz *aouls* north of the Syr-darya, and on the country watered by the Chuy, the Sary-su, and the Talass rivers.

The population is estimated to amount to 300,000 souls, of which two-thirds consist of Uzbeks and of Tajiks, who are the descendants of the ancient population of the country. Manumitted Persian and other slaves make up another 50,000, while the remainder is composed of semi-sedentary Turkomans, who occupy cultivated lands in the Khanate, or who nomadize about its western limits.

As for the last, and for the three or four hundred thousand Turkomans who occupy the Khwarezmian deserts up to the Persian and Afghan frontiers, it is evident that there is not only the land, but also the water available which could render that land fertile, and afford these nomads the means of becoming a sedentary population. At the present moment Russia is, probably, spending as much money in military establishments for overawing the wandering tribes and for preventing their taking to their former industry of man-stealing as would suffice to found well-organised settlements of Turkomans in the Khwarezmian wildernesses. In all history, what these sterile regions have denied to their wandering populations has been taken by force from adjacent and fertile countries; but, if humanity and civilisation be not mere empty words, it behoves Russia to reclaim these deserts, the saltness of whose wells fills the hearts of the Turkomans with anger and malice against all men.

# CHAPTER XVIII.

## THE LAST CHANGE IN THE OXUS.

The old beds of Doudon and Kunya-darya-lik—The Uzboy Channe of the Oxus—Arzass, Amú, and Oguz of the Great Russian Map—The dams on the old courses—Improbability of the change of the Amú having been caused by these dams—Evidence from Abulghazí Khan—Details of last change in flow of Amú, from Anthony Jenkinson and Abulghazí Khan—River enters Aral, some years after 1604—Presumption that present Amú-darya was formerly an irrigation canal, and that Aral was filled up by the waters of the Oxus subsequent to Jenkinson's day.

AMONG the old channels which formerly carried the waters of the Amú-darya, two on the left bank have been mentioned, whose dimensions are sufficiently large to allow of their having anciently been main courses of the river. One of these is a channel called Doudon, whose head was situated at about one hundred and forty miles above Khodjeili, and which, passing from thence a little to the north of Khiva, runs westwards towards the furrow at the foot of Ust Urt. It does not appear that this channel has yet been thoroughly examined, and it is believed that its traces are lost in the desert before reaching an outlet in Lake Sarakamish. The second old channel leaves the Amú, about fifteen miles lower than Doudon, and runs in a north-westerly direction towards Kunya Urgenj, the old capital of the Khivan Khanate, under the name of the Kunya-darya-lik, or 'old-little-river.' Inclining westwards from that place it curves round to the south of the counterfort, which runs out from Ust Urt in a south-easterly direction; and, thus avoiding the basin of Lake Aral, continues to the Caspian under the name of the Uzboy. From the neighbourhood of Abougir to Igdy wells the Uzboy channel has a length of about two hundred and fifty miles in a south-south-westerly direction, and forms the drainage channel into which, as has already been described, the surplus waters of the irrigation canals of Khiva have a tendency to flow.

Though Uzboy has at different epochs carried a portion of the stream of the Oxus to the Caspian, and has, therefore, been styled an ancient course of the river, it should, perhaps, be more properly regarded as the channel of that other river into which the southerly overflow from Lake Aral finally resolved itself, and which is mentioned by Masúdí in the tenth century. At a still more recent date, Uzboy may also have served as an escape channel for the overflowing Aral waters; for Russian information on the Great Map assigned to the sixteenth century, tells us that from the Blue Sea (*i.e.* Aral), the river Arzass flowed to the Sea of Khwalim (*i.e.* Caspian), and from the east into Arzass fell the river Amú, whose length was about three hundred miles. Now the Amú in this passage is in all probability the old Doudon arm of the river, and from the length given to it, its commencement would have been near Amol, a city which formerly stood in the neighbourhood of the present Tchardjui, on the western bank of the Jihún upon the road to Bokhara. This ancient city is called Amol-ul-shatt (*i.e.* on the river) by Abulfeda, to distinguish it from Amol in Mazanderan. The Amú, therefore, that flowed into Arzass, seems to have been a channel taking its name from the city, and running from the river Oguz (*i.e.* Oxus), which according to the above-mentioned Russian Map, rose in Lake Oguz (Ox in Russian), and after passing at a distance of one hundred and fifty miles from Bokhara, ran by a course about eighteen hundred miles long into the sea Khwalim (Caspian).

The lengths and directions here given to the old

Caspian branch of the Oxus, and to the Doudon arm of the Amú, are, all things considered, wonderfully accurate; but a difficulty remains to be cleared up regarding the length of Arzass, which is stated to be about nine hundred miles, while the actual length of Uzboy, from the head of Abougir to the Caspian, is about half that distance only. The obscurity as to this point may, perhaps, arise from the Arzass of the Russian Map having been measured from near Otrar, and having thus included the south-westerly branch of the Syr-darya which carried the waters of the Sihún to a junction with those of the Jihún, at the end of the fourteenth century.

Subsequently when the Amú and the Syr were tributaries once more of the Aral basin, an overflow from the Lake may again have taken place into that part of Arzass which is now called Uzboy. We know in fact, that in 1640, the two Englishmen, Reynolds and Hogg, found (early in September, before the flood season had ended, and when Aral would be nearly fullest) that the country about the head of Abougir was in a swampy state. They were told that in this locality was the old bed of the Oxus which flowed to the Caspian, and so it is not improbable that a short time previously, when the rivers feeding Åral were carrying a *maximum* discharge, the Lake might actually have been in a state of overflow. Now, it is singular that Jonas Hanway, writing in 1743, says that according to the tradition of the inhabitants of Balkhán Bay, the Oxus had ceased discharging into the Caspian about one hundred years previously, *i.e.* in 1643, or nearly the date of the journey made by the two above-named Englishmen. At this epoch, however, we know, undoubtedly, that the Amú had its mouth in Aral, and the inference is, therefore, very strong that any large stream entering the Caspian by Uzboy, in the seventeenth century, would have been due to an overflow from the Lake by the head of Abougir.

Uzboy is described as presenting the appearance of a great ruined ravine, whose sides have fallen in at intervals; and along its course and in its neighbourhood are found lakes and springs more or less salt, besides stretches of deep sand. At some places there are, however, wells of drinkable water, near which the Turkomans encamp in spring, when the country is covered with a good deal of young grass and brushwood.

On their passage through the cultivated Khivan oasis the half-filled river-beds of Doudon and Kunya-darya-lik are formed of a continuous low level chain of hollows, which are utilised as escapes for the surplus water of the irrigation canals. In an extreme high flood of the Amú there would consequently be a tendency for a considerable stream of water to flow down these channels towards the Caspian ; and, to prevent this occurring, a number of dams have been erected at various points upon their courses. To the construction of these has sometimes been attributed the change of the flow of the Amú from its Caspian to its Aral outlet; but the idea, however plausible it may appear at first sight, scarcely stands the test of close examination.

From very early times the Tartars of the Volga, true to the instincts of their descent from the wandering robbers of Asia, occupied themselves in predatory incursions upon the neighbouring countries. Masudí, the Arabian traveller and author of the tenth century, speaks of the ravages committed upon the shores of the Caspian by such freebooters, and at Strabal, in Western Khorassan, there is an artificial mound, crowned by a stronghold that was held by the Tartars of Stenkorasin, the Russian pirate of the seventeenth century. In those early days the Khanate of Khiva frequently formed the object of such attacks, and Abulghazí Khan relates in his memoirs a raid which was made upon that territory by a body of one thousand Cossacks in the sixteenth century. In the absence of the fighting-men from Kunya Urgenj the Cossacks took and plundered the city, and retired with many prisoners, both men and women. They were, however, intercepted on their retreat by cavalry, and after some days' fighting were destroyed to a man; 'since,' argued the Uzbeks, 'if we allow the Cossacks to gain the Caspian shore they will fortify themselves there, and, having received reinforcements, will eventually return and destroy us.'

Abulghazí Khan, in his 'History of the Mongols and the Tartars,'[1] has also two other passages relating to forays of the same kind, and the frequency of these attempts, combined with the dread of reprisals from Russia for attacks made in later days upon caravans, and upon the Kirghiz of the steppes, rendered, no doubt, a state of isolation very desirable for the Uzbek Khanate. General Ivanien, in his pamphlet on Khiva, relates the following circum-

[1] Translation by Baron Desmaisons: St. Petersburg, 1874.

stance, which places such apprehensions in a striking light:—A Russian prisoner in the town of Gorlen in 1834 described the Amú flood of that year to have been so high that a large volume of water escaped down the Kunya-darya-lik into Uzboy, and the Khivans sent men specially to see to what distance downwards the water had reached. 'For,' said the Uzbeks, 'Russia is building a fortress (Nova Alexandroskiya, on the Caspian shore) upon our soil, and the river is going down to meet them.'

The existence of the dams on the Kunya-darya-lik lends, therefore, some air of plausibility to the idea that the flow of the Amú was artificially changed on account of the political exigencies of Khiva, as understood by her rulers. These dams could scarcely, however, have been built until the stream of the river had declared a decided inclination to flow by its northern course to Lake Aral. Nor is it probable that Abulghazí, a Khan of Khiva himself, would have omitted in a work relating to his family and his country all reference to a state policy which, if it ever existed at all, must have been well known to him, and was actually consummated in the time of his father. His silence on the point affords a very strong presumption that the last change of the Amú from its Caspian to its Aral outlet was actually due to those phenomena which are characteristic of the river, and which have been described on previous pages. And an examination of such records as are attainable regarding this change will still further confirm this theory. In 1559 Anthony Jenkinson, an Englishman, on a commercial mission to Bokhara, travelled from the north-east coast of the Caspian and passed through Khiva; and the information he recorded[1] regarding the Oxus of his day, taken in conjunction with passages in the 'History of the Mongols and the Tartars,' enables the whole process of the last change in the flow of the Amú to be traced in a tolerably complete manner.

In a passage of the work last mentioned, we are told, that in the commencement of the sixteenth century, the territory of the Khans of Urgenj still included the country down to the south-east of the Caspian; and that Dín Muhammed Sultán, when going from Kunya Urgenj to Mazanderan, followed the banks of the Amú to Tchikdelik (Igdy), where

[1] Jenkinson's journey is to be found in 'The Principal Navigations, Traffiques, and Discoveries of the English Nation, made by Sea or over Land.' By Richard Hakluyt, Preacher and some time Student of Christ Church, Oxford. Imprinted Anno 1599.

the river makes an elbow, and from whence he passed southwards to the wells at Dinár. Abulghazí also tells us that at the same date, all the road from Kunya Urgenj to the Balkhán mountains on the south-east coast of the Caspian was covered with *aouls*; for the Amú, after passing the city named, flowed down to an outlet at Ogourtcha (in Balkhán Bay), and both banks of the river to its mouth were cultivated and covered with vineyards and orchards, and were very populous and in a most flourishing condition. From these passages, therefore, we learn, that early in the sixteenth century some of the Amú waters flowed by the Kunya-darya-lik into the Uzboy channel, and so past Igdy wells to the Caspian.

After twenty days' journey, from the east coast of the Caspian, through a waterless desert, Jenkinson says he arrived, early in October 1559, at 'a gulfe of the Caspian sea againe, where we found the water very fresh and swete. . . . Note that in times past there did fal into this gulfe the great river Oxus, which hath his spring in the mountains of Parapomisus in India, and now cometh not so far, but falleth into another river called Ardok, which runneth toward the north and consumeth himself in the ground, passing underground above 500 mile, and then issueth out again and falleth into the Lake of Kitay. We, having refreshed ourselves at the foresaide gulfe, departed thence the 4 day of October and the seventh day arrived at a castle called Sellizure. . . . The castle of Sellizure is situated upon a high hill, where the king called the Can lyeth. . . . The south part of this castle is lowe lande, but very fruitfull, where growe many good fruites. . . . The water that serveth all that country is drawen by ditches out of the river Oxus, into the great distruction of the said river, for which cause it falleth not into the Caspian Sea as it hath done in times past, and in short time all that land is like to be distroied, and to become a wilderness for want of water when the river of Oxus shall faile. . . . The 14 day of the month we departed from the Castle of Sellizure, and the 16 of the same we arrived at a citie called Urgence.'

It was, then, three days' journey from Kunya Urgenj to Sellizure, and the same from Sellizure to the 'gulfe of the Caspian,' which was, therefore, six days' journey west of Kunya Urgenj, or, say, one hundred miles in a direct line. This would put Jenkinson's 'gulfe of the Caspian,' at the lake Sarakamish, into which water would pass by the

Kunya-darya-lik channel of the Amú, on its way down to the Caspian. Jenkinson was told that the river flowed here formerly (information which corroborates Abulghazi's description of the course of the Amú), but he found that it stopped short a little distance of Sarakamish, since it reached as far (at least) as Sellizure, three days' journey east from the 'gulfe of the Caspian.' The water of this place is stated to have been very 'fresh and swete' by Jenkinson in October, and it must certainly have been due to the yearly summer floods of the river, which had recently subsided. It has already been stated that Sarakamish occupies an indentation or bay in the limiting cliff on the south-east of the Ust-Urt plateau; and would have thus been a 'gulfe of the Caspian Sea' at a time when the Aral was connected with this basin by an overflow of its waters. It may be inferred, therefore, that the people who informed Jenkinson that the place in question was a 'gulfe of the Caspian Sea,' preserved the tradition of the former physical aspects of the locality.

To Sellizure, three days' journey west of Kunya Urgenj, the Amú, therefore, still flowed in October 1559, though there were great apprehensions that the stream would soon fail, or, in other words, would cease to flow so far; and this fear was soon to be realised, as we find from a passage in Abulghazí Khan's memoirs. He tells us that he was born in the year '1014 of the Hijra, in the year of the Hare, when the sun was in the sign of the Lion, on a Monday at daybreak, the 15th of the Rubbee-ul-awal'—or, in other words, his birthday was in July 1605—and that thirty years previously to this event, or in 1575, 'the Amú-darya cut for itself a channel from the place called Kara Oighour Tokai, above Khást Minár, and, taking the direction of Tuk-kala, passed on towards the Sea of Sir (Lake Aral)—a circumstance which changed the neighbourhood of Kunya Urgenj into a desert,' *i.e.* by depriving it of the water necessary for its fertilisation. The ruins of Tuk-kala have already been mentioned as still existent on the bank of the Chertambye channel, near where this leaves the Amú about twelve miles below Khodjeili; and there is consequently but little doubt that the channel mentioned in the above passage is the present course of the Amú at the place specified. Another passage of Abulghazí's work tells us that a year before his birth, *i.e.* in 1604, his father caused a canal to be dug which commenced from above the fortress of Tuk, and 'that some years later this canal had a breadth of more than an arrow's flight. The water passed by Quïghún, from whence it fell into the sea.' Here we see that this artificial canal gradually enlarged itself and fell into the Lake Aral some years subsequent to 1604, while the flow of the Amú to Kunya Urgenj stopped in 1575, and would, therefore, have ceased a few years previously to reach Sellizure, a distance of three days' journey further to the west. This information is entirely in accordance with that of Jenkinson, who stated in 1559, that the Oxus was likely to fail at the place in question.

The exceptionally gentle slope of the country between the head of Abougir and Igdy wells, as compared with that of the other localities, watered by streams from the Amú, necessarily invites a great deposition of sand in Uzboy, and so forbids any permanency of the river's flow down this old channel of the Oxus. However, in a high flood some portion of the waters of the river might still, perhaps, pass down by Kunya-darya-lik or Doudon into Uzboy, and so on to an outlet in the Balkhán Bay. As has been seen, there was actually some apprehension of this having occurred so late as in 1834, and even up to 1857 (the Turkoman chief Atta Murád Khan told me), water entered Lake Sarakamish; though such found its way from the Amú to the place in question by the Loudon canal, which detached a branch into the Kunya-darya-lik.

If we refer again to Jenkinson's narrative, further proof will be found that the last change which took place in the flow of the Amú-darya, was due to the action of those natural causes whose working is, at the present day, so observable in the channel of the river. 'The 26 day of November,' says Jenkinson, 'we departed from the town of Urgence, and having travailed by the river Oxus one hundred mile we passed over another great river called Ardok . . . . This river Ardok is great and very swifte, falling out of the foresaide Oxus, and passing about 1,000 mile to the northward, it then consumeth itself in the ground, and passing under the same about 500 mile, issueth out againe and falleth into the lake of Kitay, as I have before declared. The 7 day of December following, we arrived at a castle called Kât.'

From the upper end of the Kunya-darya-lik to Urgenj is very little more than one hundred miles, and Kât, the castle reached by Jenkinson in thirteen days' journey from Urgence, was situated (as has been recently shown by the researches of Russian officers in 1874) near Shahbázwálí, on the right bank of the Amú, nearly opposite to the head of the old Kunya-darya-lik channel.[1] The present course of the Amú, consequently corresponds with that of Ardok, which, being a 'greate and very swifte' river, shows that in 1559 its channel was not large enough for the volume of water which was being thrown into it by the gradual obstruction of the Kunya-darya-lik bed. How small it actually was, may also be gathered from the consideration that in the month of December the volume flowing in the Amú would be approaching its minimum, as the sources of the river would have been icebound, and that some portion of this small winter stream was flowing in the Kunya-darya-lik bed of the Oxus at the moment. It may therefore be safely inferred that Ardok was originally nothing more than an artificial irrigation canal, running towards Khodjeili. On reaching that place it seems tolerably certain that its waters passed to the north-east by the Karauzak channel (which has been previously mentioned), into the low ground situated near Lake Dowkara on the south of

[1] Kât was a city of Khwarezm, stated by Oriental authors to be on 'the eastern parts of the Jyhoon or Oxus, nor was there any other eastward of it in the districts of Khwarezm,' and it, therefore, would be passed by Jenkinson on his way to Bokhara.

the Belitao hills; since we know, from the passages which have already been quoted from Abulghazí Khan, that the course of the river westerly from Khodjeili was an artificial canal dug some years subsequent to 1559.

As Jenkinson proceeded directly eastwards from Kât to Bokhara, it is evident he had no personal knowledge of the outlet of Ardok, and that his wonderful statements on this head are merely grounded on hearsay. Whatever truth, however, there may be in these, points to the conclusion that Ardok had no defined mouth in Lake Aral; a circumstance which is directly confirmed by Abulghazí's information that the river did not enter that sheet of water till some years subsequent to 1604. Ortelius' map, it has already been said, shows the Syr as flowing into the Kitay Lake, from which the River Obi rose; and it may, therefore, be that the dried-up state of the Aral basin in the 15th century was preserved during the 16th, owing to the change in the flow of the Syr out of the Uzboy channel into the basin occupied by the Chalkar-tangiz and Sary-kupa lakes, on the north of the present water-spread of Aral. The waters of Ardok would subsequently have found their way into the Aral basin proper and have gradually filled it up, until the Sea of Sir, which Abulghazí Khan speaks of, was formed to receive the whole discharge of the Oxus, when the river abandoned its Caspian outlet and transferred its flow entirely to the more elevated basin.

# CHAPTER XIX.

## RUSSIA ON THE AMÚ.

The village of Núkús and the new Russian fort—Petro Alexandroskiya—Military life on the Amú—The new Russian territory—Population—Revenue—Possible improvement—Trade—Old Russian slaves—The Amú as a navigable stream—The Yany Su and Kuwán-jerma—The upper courses—Descent of river by a steamer—Native boats on the Amú—Isolation of Amú district from the Syr-darya—Project of filling the Jani-darya—Objections to such an enterprise.

JUST above where the Kuwán-jerma arm leaves the Amú-darya is situated the small mud-walled village of Núkús on the right bank of the river, which is here crossed by a ferry, made much use of by the scattered population, converging upon Khodjeili on market days. A mile or two north of Núkús is the new Russian fort—a small square, enclosed with *pisé* walls and flanked by two circular bastions at opposite angles. It contains barracks for its garrison of two companies of infantry, who will, with the armament of four field-guns, be more than a match for any number of thousands of possible attacking Turkomans or Uzbeks.

The surrounding flat country is slightly cultivated, but is for the most part covered with a tamarisk and *eleagnus* jungle, which is thick and rather large along the dry course of Karauzak. Hereabouts in 1873, after the close of the Khivan campaign, a Russian camp was established on the banks of a stream flowing from the Amú into Kuwán-jerma, and the troops were employed in the construction of the fort. This military post forms a link between the base of Lake Aral and the advanced position of Petro-Alexandroskiya, which is situated some two hundred miles further up the river on the same bank.

Petro-Alexandroskiya is the chief Russian post in the Amú-darya district, and its garrison of some two thousand men are lodged in a large fortified garden, which was the property of a Khivan magnate. It is two or three miles from the right bank of the Amú, in about the same latitude as Khiva, from which it is about thirty miles eastward. A small town of the usual Central Asian steppe pattern is springing up at a little distance from the encircling wall and parapet of the garden, which has been acquired by irrigation from the adjacent desert. The verdure and leafiness, as well as the human movement of the place, form a pleasing contrast with the sterility and desolation of the sands which lie at its very postern gates.

Notwithstanding the constant drill and rifle practice, the Russian officers on the Amú would find life dreary were it not for the occasional brush with the Turkoman tribes, which promises to become a periodical institution. Cut off as they are by eight days' post from Cazalinsk, which in its turn is separated by three weeks' post from Moscow, the Europeans in this part of Turkestan are placed in a disadvantageous position as regards a regular supply of literature. Besides a little tiger slaying or pheasant and snipe shooting, means of recreation are totally wanting, and such an isolated position, besides aiding, perhaps, to a caducity of European *morale*,[1] is in itself an incentive to that military restlessness and yearning for decorations which is taken as a sign of the aggressive policy of Russia in Central Asia.

[1] Infandum! adolescentes Bokharæ, urbe stuprorum, lupanari; palam ac jure venundari apud Vamberium narrantur. Utrum censorum Scythici istæc permittent nescio, fœdissima percontari non soleo, confiteor; attamen credo hæc probra etiamnum florere, nam greges

Though the hospitality met with in Russia by a stranger is proverbial, the great cities of most countries are not very different from one another in this respect. It is in the standing camps of Turkestan, or still more, perhaps, in the desert bivouacs, that the geniality of Russian officers impresses one as being not the least of their military virtues. The anniversary *fête* of a regiment is a good opportunity for observing this quality, and churlish indeed must that guest be who is inaccessible to the kindly, if somewhat boisterous, attentions of which he is the object.

The boundary of the territory of the Russians on the Amú is the main course of the river, from the Taldyk mouth in Aral to Meshekli on the new Bokharan frontier—a distance of about two hundred miles above Khodjeili. At the southern end of this line the eastern desert extends up to the low hills which skirt the river for a distance of thirty miles downwards; but, in the next eighty miles below, a strip of country, about five miles in

adolescentium ex intima parte Asiæ Mediæ sub tutelâ et cum assensu censorum, agrum Scythicum stupri causâ perlustrare solent. A.D. MDCCCLXXIV. grex impuberum in castris Nukii et Cimbæ (in regione Oxii) lubricas saltationes Scythicas centurionibus gratissimas exhibebant. Prudenti satis.

breadth, has been reclaimed by irrigation from the surrounding sands on the right bank. Excepting the half-drowned lands of the Karakalpaks, this area of, perhaps, four hundred square miles is the only productive part of the Russian possessions. Though it has been stated at a higher figure, the population of the Amú-darya district, including Karakalpaks, Kirghiz, Uzbeks, and Tajiks, does not amount to more than 60,000 souls.

The revenue collected in 1874, the first year of the Russian occupation, amounted, as far as can be ascertained, to some 12,000 roubles (say 1,850*l.*)—a sum which was about one-eighth of the expenditure. The assessment on the land has been reduced to less than that demanded in Khiva,[1] with the view of attracting an influx of settlers; though it is doubtful whether the physical features of the ground will allow of any appreciable addition to the irrigated area on the Russian side of the stream. It is, however, evident that any measures which should divert water for the cultivation of the desert, and so relieve the lands of the Karakalpaks of the Lower

[1] The highest land tax in Khiva is about two shillings per acre; the lowest about six pence. On the Russian bank of the Amú, the State demands a rent of one-tenth only of the produce of its own property from tenants.

Amú from being inundated, would add to the revenue in a twofold degree. The encouragement of cotton cultivation on the waste lands surrounding the Khivan oasis would greatly benefit Russia, who at present pays eleven millions of roubles (say 1,600,000*l.*) yearly for a very insufficient supply of raw cotton. There is, perhaps, little doubt but that attention generally would be drawn to such a scheme, were it not for the existing difficulties regarding the transport of goods from Turkestan into Europe. A future reclamation of the desert would in any case be advantageous, as it would provide for the settlement of the nomadic Turkoman tribes, whose marauding propensities can only be successfully modified by giving them lands they could cultivate.

The commercial spirit which, doubtless, exists in these regions promises to receive some development, for the Russian occupation has commenced to attract a floating population from Bokhara and from the upper courses of the river for purposes of trade. Caubul merchants would, probably, find a lucrative market for Himalayan teas, since the very commonest kind used by Russian soldiers costs 1¼ roubles per pound on the Amú. After all that has been said about Uzbek intolerance, it is remarkable to note the comparative freedom which Russian traders have enjoyed for a long time in the Central Asian Khanates, and how many slaves of that nationality have succeeded in making themselves good positions—in Khiva more especially. In the old days the price of a Russian was two hundred tillas (about 60*l.*)—a sum which was three times that paid for a male, and four times the price of a female Persian. These latter always remained slaves, though their descendants in the second generation became free; but a Russian usually purchased his freedom out of his savings, and very frequently settled down in the Khanate. There are many examples of such, who have risen to influential positions; and one is to be found in the person of an aged Russian who is a respected local magnate at Shahbázwálí. He has married a Mohammedan woman, and has a family of children, but has himself remained an orthodox Greek in religion.

The analysis which has been given of the phenomena attending the flow of the Amú is of itself sufficient to show that the river can never make a convenient line of water communication for navigation by steamers, though if the gap at Toyuboyin were once passed through there would, probably, be no obstacle found to the ascent of steamers five hundred miles farther, to the frontiers of Afghanistan.

The Russian steamer 'Samarcand' has already been described as having ascended the Ulkun or central branch of the lower Amú, whence she passed through the lake region to an anchorage under the plateau of Kashkanatao. From this locality troops and stores are despatched to the upper courses of the river, on native *caïques* of large size, by the Chertambye channel, and an unsuccessful attempt was made in the summer of 1874, by the steamer 'Peroffsky,' to follow the same route. She subsequently passed out of the river into Lake Aral, and achieved her passage to the main Amú, through the eastern arm of the river. Though the depth in this is scarcely more than just sufficient for a steamer, except in the flood season, such an objection is of no great importance, as the Aral flotilla vessels are not on duty in the winter. In the summer the velocity of the stream, and the numerous bends in Kuwán-jerma, throw considerable difficulties in the way of its navigation, and point to the desirability of a special naval establishment, which shall be affected to the use of the Amú.

In the upper courses of the river the whole engine-power of the 'Peroffsky' was frequently neutralised in contending against the strong current, and, probably, sixty per cent. of the whole quantity of fuel she used was absorbed on this account. Her firing cost about 2*l.* 5*s.* a mile for the one hundred and forty miles of the river which she ascended from Khodjeili, but on this, the first occasion when a steamer was seen on the ancient Oxus, time and money might have been saved had proper preparations been made. The local impression was that the 'Peroffsky' would not succeed in ascending the eastern arm of the Amú, and consequently any fuel procurable, such as poplar-poles, &c., however unsuitable it was for firing, had to be bought at various places during the voyage.

The descent of the river was a far more disquieting, if it was a shorter, enterprise than the ascent, for the chief use made of the 'Peroffsky's' engines was to go astern, in order to avoid running, with the current, stem-on to the numerous shoals, which had been deposited in the channel since her passage, some ten days previously. It says a good deal for the goodness of the bargain made by the Russian Government, in the purchase of the 'Peroffsky,' that, though she has been frequently running aground in the Syr-darya for the last twenty years, she suffered little or no material damage from the many mishaps of the same kind which she met with in 1874 on the Amú. The numerous shocks which her machinery underwent unharmed, in her frequent collisions with sandbanks, are worth volumes of testimonials in favour of the firm of Swedish engineers who constructed the engines of this steamer.

The navigation carried on by native boats upon the Amú and its various branches is admirably adapted for all local demands of trade. The *caïques* vary in size, from the little canoe of the solitary Uzbek fisherman to the large transport barge of fifty feet long and eight to ten feet broad, constructed of three-inch poplar or elm (*karagatch*) planks, which are fastened by iron clamps. With a loaded barge it is a work of difficulty to ascend the stream, and the operation is effected by the crew (amounting to half-a-dozen in large boats) towing it by a long rope, to which they are harnessed. In descending, the boat floats with the current, and is kept in the full stream by an occasional stroke of an oar, or shove of a pole, which is handled by one of the men, while the remainder are sleeping. For merchandise, prices range from a halfpenny per ton per mile down-stream to three times that rate up-stream, while for passengers it is more than double.

In speculating on the causes of the small importance of Asiatic rivers as commercial routes in antique times, Heeren has suggested that the scarcity of timber and, in many places, of iron prevented that development of water communication which has been attained upon the streams of Europe. Without questioning this explanation, it is evident that the hydraulic character of the Oxus, and indeed of all rivers descending from the Central Asian highland, however favourable such may be for downstream summer navigation, offer great impediments to their being navigated up-stream. Still more, the changes of the courses of these rivers, that are brought about by the practice of irrigation, have not allowed of the permanency of position for the terminal depôts of trade, in which goods might be exchanged, or might be stored, before being forwarded to the consumers. Capital directed by engineering knowledge can surely overcome these natural difficulties at the present day; but their absence during antique times throws much light on what has been hard to explain while our knowledge of these rivers was less complete than it now is.

Isolated, as the Russian district of the Amú-darya is, from its support on the Syr-darya, by the Kizzelkoom desert and by the inconvenient approach from Lake Aral, it is natural that attempts should be made to overcome the existing difficulties of communication; and the means for effecting this which are apparently under consideration, is the diversion of a portion of the stream of the Syr, by the Jáni-darya bed, to the south-west. The country along this line will be presently described in detail, but what has been already said of the rivers of West Turkestan will indicate that the permanent flow of water in the Jáni-darya cannot be ensured. As a mere temporary measure such an enterprise would, of course, provide a line of communication across the wastes, between Fort Peroffsky and the Amú; but its accomplishment would probably deprive Cazalinsk and its neighbourhood of the water necessary to prevent the country from becoming desert.

In such a case the difficulties and length of the present route between Europe and Asia would be added to, since Fort Peroffsky would replace Cazalinsk as the first habitable station of the possessions of Russia in Turkestan.

# CHAPTER XX.

## THE BANKS OF THE AMÚ.

The river from Khodjeili to Bend—The Bishtubye ridge—Syed-atta-auliya—Nogai-kala—View of Kizzelkoom—Miskán-atta—Nazr Khan—Tower of the Jinn—Kipchak—College of Niyázbye—Instances of the crossing of the Amú on the ice, recently and in old times—The Governor of Kipchak—His Kirghiz attendant—Ruling minorities and ruled majorities.

ABOUT the village of Núkús, whose walls give it quite a fortress-like appearance, there is some little cultivation and timber, but less than on the opposite Khivan bank of the Amú, where thick clumps and groves of trees entirely hide the houses of Khodjeili. The environs of this town, however, soon become open, and a few scattered homesteads are seen among the crops, which give place farther south to the *eleagnus* jungle, skirting both banks of the river generally, along forty miles, up to Kipchak.

Between Khodjeili and Bend, a distance of less than twenty miles, the breadth of the river varies but little, and in this portion two or three irrigation canals are diverted for the irrigation of land on the Khivan side. Their heads are concealed by low alluvial islands covered with jungle and high grass; and the channel of the Amú, which broadens at such points, narrows again immediately below them and so increases the velocity of the stream.

Bend, a fortress erected for the protection of the dam thrown across the Loudon canal, is hidden in the jungle, which is here rather higher and thicker, and through which a small stream of water still flows from the river, in a north-westerly direction, to fertilise a limited area of land, between Kunya Urgenj, and the head of Abougir.

The opposite or right bank is occupied by the bare Bishtubye ridge, averaging, perhaps, two hundred feet in height, which commences a mile or two above Núkús, and skirts the river more or less closely for fifty miles upwards, to the Shaikhjaili hills. The formation of Bishtubye is a reddish-brown indurated clay, precisely like the other elevations of the lower Amú, which have been already described; a few higher points occur along this ridge, and these are capped with strata of ferruginous sandstone. The northern extremity of Bishtubye is occupied by a hill called Syed-atta-auliya, that is crowned by a mausoleum, whose surrounding cemetery presents a curious appearance, from the numerous wooden stretchers, which are used in carrying corpses, being planted vertically in the ground close to the graves.

The whole way down from the Shaikhjaili hills the encroachment of the Amú on its right bank is very noticeable, and above Syed-atta some of the low ground between Bishtubye and the river is being changed into marsh, and will soon be absorbed into the channel. Opposite to Bend there is an elevation on the ridge, called Nogai-kala, from the circumstance of some Nogai Tartars having fled here after the taking of Kazan, by John the Terrible. From its summit glaring sandy wastes are seen extending to the eastern horizon, but the eye finds relief to the south, where a sheet of verdure lies between the ridge and the edge of the river. This belt of jungle vegetation, full of lagoons fed by the Amú, is no more than a mile or so in width, and is thus scarcely worthy of being dignified by the name of a forest, which Vámbéry applied to it. Nor are the panthers, lions, and tigers, as far as can be learnt, so numerous as that traveller's informants led him to believe they were; though, on the other hand, not a few cattle find tolerable pasture in this jungle, up to Nazr Khan, about thirty miles above Núkús.

Under the edges of the strata of ferruginous sandstone which cap the summit of Nogai-kala are some caves frequented by owls of a very large size; and the unwary visitor, whose curiosity may be excited by these excavations, is not unlikely to be upset and rolled down the steep slopes of the hill, by the sudden flying out of the birds on his approach. At the foot of Nogai-kala an open plain of more than a mile in breadth stretches down to the river's bank, and on it a great many fossils are found, including ammonites of about eighteen inches diameter. These gigantic shells are firmly embedded in the indurated clay soil, which atmospheric action has worn away slightly around them, and so gives them the appearance of the bases of ruined pillars, and has probably given rise to a local tradition that the place is the site of an ancient city.

On the opposite, or Khivan, bank of the Amú the *eleagnus* jungle continues, for some ten miles from Bend, to a rising ground which has been cut away steeply by the stream, and on whose edge is a small mud fort; behind which the ground rises gently inland to some tombs surrounding the mausoleum of Miskán-atta. Such provincial monuments must not be imagined to possess any architectural features, for they are usually mere earthen platforms, forming the base of a small domed structure in *pisé*, which is decorated with rags suspended around it by pious pilgrims. It is only in towns of the Khanate that the mosques containing the tombs of venerated saints, such as Polwan-atta, are built of burnt brick and ornamented with bands of artistically-enamelled tiles, produced by the characteristic manufacture of Khiva.

The Amú, which has from Bend to Miskán-atta the considerable breadth of nearly a mile, owing to the former diversion of its waters by the Loudon canal, contracts again near Nazr Khan. At this place the marshy jungle under the Bishtubye ridge opens out, and is occupied by a large homestead, very similar to those previously described, and which is the property of a Kirghiz chief, whose ancestors, more than a century ago, possessed themselves for a short time of the throne of Khiva. The submission then made by them to the Czar, is held to establish the quasi-legal suzerainty of Russia over the Khanate at the present day.

The banks of the Amú-darya up to Nazr Khan can scarcely be considered either interesting or picturesque; for the bare clay slopes of Bishtubye, and the muddy stream fringed with low jungle, possess an air of monotonous solitude, that is but occasionally relieved by the small *caïque* of an Uzbek fisherman. At Nazr Khan, however, some improvement takes place, for to the south the broken outlines of the Shaikhjaili range begin to declare themselves plainly, while in the middle distance the low, massive Tower of the Jinn crowns a conical hill lying on the plain between the river and Bishtubye, which is here at a greater distance from the Amú. Inside the walls of the frowning tower is a deep well, which figures in the local romance of a princess, who, when besieged, in company with one of her father's slaves of whom she was enamoured, caused a subterranean passage to be dug down to the neighbouring stream of the Amú.

From Nazr Khan, upwards to the Tower of the Jinn, the jungle covering the right bank increases in size, while the low Khivan bank opposite commences to be clothed with cultivation, that is watered by canals from the river. The fields are interspersed with timber, which grows thicker and thicker, until the pretty groves in the environs of Kipchak are reached, and give the first earnest of that famous oasis which the ascent of the Amú-darya, so far, has failed to discover to the curiosity of the traveller.

The gardens and clumps of trees extend along the Amú, from a point opposite to the Tower of the Jinn, up to Kipchak, a town of mud houses, whose surrounding high wall runs along a sort of natural quay upon the river's bank, which is covered with piles of timber and stacks of agricultural produce, and has some boats moored alongside of it. On the opposite bank is an open plain, where stands the college erected and endowed by Hajjí Niyázbye, a native of Kipchak, for the support of some twoscore theological students. This building is one of the few good specimens of Mohammedan architecture to be seen on the river, and is built of burnt bricks, each one of which, local report says, cost about sixpence—which is not an improbable circumstance, when the scarcity of firewood in these regions is considered. The college has an arcaded front in two storeys, about two hundred feet in length, and fifty feet high; and close to it are situated some handsome domed mosques, containing the tombs of the founder's family.

It was near the hill on which stands the Tower of the Jinn that the Russian chief of the Amú district marched his troops across the ice in the winter of 1873, to punish the Yomut Turkomans—an operation which he repeated again in 1874, lower down the river, near Khodjeili. Abulghazí Khan describes a similar feat of arms, which took place at a point a little above Kipchak, in the winter of 1624, between the Uzbek party of Khiva, under his own command, and the Turkomans of the Caspian littoral, under his half-brother, Isfendiyar Khan. Abulghazí, having the smaller force, entrenched himself across Bishtubye, with one flank resting on the river, and the other covered by a line of baggage-carts, and awaited the Turkomans, who crossed the Amú on the ice.

'The sun,' says the Khan, 'was already two lance-lengths above the horizon when the enemy came up and put himself in battle order, upon foot as we were. I myself counted his numbers, which were five thousand, and our own, which were four hundred. In the battle which ensued, and which lasted from one prayer-time to another, the Uzbeks lost twenty men killed and one hundred wounded, while the enemy had seventy to eighty killed and two hundred wounded. He spent the night on the battle-field, and retreated at the following day-break.' Making allowances for all probable exaggerations, the passage conveys an idea of fighting which is far above the standard of the degenerate Uzbeks of the present time.

When the 'Peroffsky' arrived at Kipchak the governor of the place came off with his suite to inspect the steamer, and to make arrangements for a supply of fuel to replenish her empty bunkers. He was a frail-looking Uzbek, whose health seemed to demand a tonic in the shape of *vodka*, of which liquor he took away several bottles on his departure. Among his followers was a gaunt, ugly, swivel-eyed Kirghiz, on whose senses the tea which he drank on board the 'Peroffsky' had the cheering and even exhilarating effect usually produced by whiskey-punch on an Irishman. He became excessively friendly, not to say familiar, with the Russian officers on board the steamer, and his demeanour formed a striking contrast to that of Hindús towards Englishmen, making me think that the possible dislike of a clean and well-bred fellow-subject, was infinitely preferable to the attachment of an ill-bred and dirty one.

Whatever satisfaction may be felt by Russians at the foregoing avowal, I gladly make them a free gift of; for if native opinion in India still thinks the 'barbaric pearl and gold' of a past tyranny to be a worthier thing than law and freedom and order, the error is certainly none of the Government. To the British public, who are too prone to condemn their countrymen at a distance, I would remark that since the days of Burke, down to the present time when the Heir to the Crown has devoted himself to softening the intercourse between rulers and ruled in Hindustan, little but mere brute force on the one hand, or hireling special pleading on the other, has been contributed by England for the amelioration of those antagonistic feelings which human nature unfortunately, but necessarily, engenders between a small dominant alien minority and an enormous subject majority.

# CHAPTER XXI.

## THE BANKS OF THE AMÚ—*continued.*

The Amú between Kipchak and Shaikhjaili—The Shaikhjaili hills—Action of stream on the right bank—The continuation of Bishtubye above Shaikhjaili—The Khivan oasis—The Russian bank—Shahbázwálí—Shurakhan—The river at Tunuklu—The Valley of the Lion's Mouth—Sultan Serai—The island of Aralchí-baba-auliya—Toyuboyin.

Above Kipchak the Khivan bank of the Amú continues to be timbered, but the cultivation retires a little from the river inland, and on the horizon westwards homesteads are seen dotted about the country for twenty miles upwards, in the direction of the town of Mangit. At the upper end of this distance, on the edge of the Amú, and opposite to the Shaikhjaili hills, are some detached outlying elevations of the same geological formation as this range.

The Bishtubye ridge is lost in the lower slopes of Shaikhjaili, which rise rather abruptly two or three miles east of the Amú, to which they run directly west before turning southwards to overhang the water along a distance of about eight miles. Their western face on the river presents a massive group of, perhaps, 2,000 feet high, composed of a metamorphic schist of a sombre brown colour, approaching in some places to a neutral tint, and having a greenish hue like that produced by coal-dust. The crystalline rock has the appearance of being tipped with glistening points, while larger masses of the same material crop out in irregular and tilted *strata*, which are more noticeable at the rounded shoulders of the ravines separating the buttress-like portions of the central mass. Here and there in the hills are lighter-coloured patches, whose *detritus* forms heaps at the foot of the steep slopes on the edge of the water below.

The breadth of the Amú at Kipchak is great, and the low eastern bank, on which the college built by Hajjí Niyázbye stands, is much cut away by the stream which is deflected against it from some rocks situated upon the point of land that is occupied by the town. These rocks have been mentioned by Vámbéry, but do not present so marked a feature to-day as they appear to have done at the date of his journey (1863). They were, however, sufficiently noticeable in 1874 to cause the captain of the

'Peroffsky' to give them a wide berth, and to ascend by the channel near the opposite bank of the Amú.

The river preserves its breadth up to the commencement of the Shaikhjaili range, and along the whole of this portion the stream acts strongly on the right bank, whose destruction is aided by the falling in of caverns in the clay, which are left by the washing out of the sand they contained. On the ground between the Amú and Bishtubye is the Khodjakúl lake, which will soon be absorbed into the channel, together with a small domed mosque and cemetery lying between this lake and a plantation of trees, that cover the bank of the river for some little distance upwards from a point opposite to Kipchak.

Upon the crest of the Bishtubye ridge, above Khodjakúl, stands a large stumpy minaret, forming a remarkable landmark. Its erection has been attributed to Bekovitch's expedition, and a second one is found on the high ground continuing along the bank of the Amú above the Shaikhjaili hills, in prolongation of Bishtubye; of which it seems formerly to have been a part, and from which it was apparently separated by the upheaval of those hills.

At the upper extremity of their river-face the Shaikhjaili hills turn eastward and stretch into the Kizzelkoom desert, with decreasing height along a distance of fifty miles, to their termination. Remarkable contrasts are afforded by the dark southern spurs of Shaikhjaili, which are intermingled with lighter-coloured clay and chalk ridges continuing upwards along the bank of the Amú, and give the idea of lines and clumps of massive timber. The careful use of a field-glass is required to show that all these slopes are entirely bare of vegetation, and, indeed, characterised by great sterility; and this is mentioned somewhat pointedly, as it may explain how Vámbéry came to describe the slopes of Shaikhjaili as being covered with forests.

In the reach of the Amú flowing past Shaikhjaili the channel is less than three thousand feet broad, and contracts at the upper end near the ruined fort of Yampuk, which stands on the right bank opposite to the detached elevations that have been mentioned. So far in the ascent, both banks of the river are capable of being examined from the deck of a steamer, but immediately past Yampuk, for a distance of ninety miles farther up to Tunuklu, the breadth of the Amú expands so much, and the channel winds in and out between such large alluvial islands, that either the one or the other bank only is occasionally to be seen.

Commencing from nearly the same point, nothing can be more striking than the change which has been effected by the industry of man, aided by the waters of the river, in transforming the sterile clay deserts into a smiling and highly-cultivated garden; while at the same time the want of life and movement which attaches to the Amú in its lower courses is here replaced by frequent boats carrying men or agricultural produce, or by groups of labourers along the banks. The numberless homesteads scattered over the cultivated fields are tenanted by the great mass of the population who are agriculturists, while hidden behind the leafy groves are many considerable towns inhabited by *employés* of the State, or by traders and artisans.

The right or Russian bank is far from approaching the Khivan standard of fertility, except over a very limited portion of its area; though it is redeemed, by water conveyed from the Amú, from the aspects of sterility by which it would otherwise be characterised. Most of the cultivation in Russian territory is situated some little way up the river, but the low jungly ground immediately south of Shaikhjaili affords good pasture for numerous flocks of cattle and sheep, whose shepherds, attended by crop-eared, short-tailed dogs, are here frequently seen in groups.

Though cultivation gradually replaces these pasturages of the right bank, timber is less plentiful than on the Khivan side, and the buildings which commence at Rakmanbirdibye about twenty-five miles above Shaikhjaili, are, in consequence, more easily seen. A little farther is the considerable town of Shahbázwálí, where there is a fine tall minaret of the usual Khivan design, ornamented with horizontal chequered bands of enamelled tiles, and just beyond this edifice is a group of domed mosques. Besides the town and college near Kipchak, the buildings about Shahbázwálí are the sole architectural features to be met with upon the banks of the lower Amú-darya.

The ascent of the Amú is continued for twenty-miles farther between the islands due to alluvial deposits, which have also formed the low right bank along this portion, up to a point abreast of Petro Alexandroskiya. Here the islands are covered with waving high grass, while the river's bank is clothed by a sheet of the rose-pink flowers of a thick, low growth of tamarisk bushes, beyond which is a line of green poplars hiding the fields of maize and other cultivation that surrounds Petro Alexandroskiya and Shurakhan. This latter place was visited and described by Vámbéry, and offers little interest, being but a repetition of the semi-ruinous towns of Kungrad and Cimbye. Placed, however, as it is on the very edge of the Kizzelkoom, the beautiful verdure of its irrigated fields and gardens are a striking example of the fertilising power of the Amú waters upon the clay of the Khwarezmian plain. Between Petro Alexandroskiya and Shurakhan, a distance of some four miles, the cultivated tract is crossed by narrow strips of desert running westwards and presenting the appearance of abandoned land which has been cultivated at a former date by canals flowing from the east; thus suggesting the idea that they must have carried the waters of the Amú when that river had a more direct northerly course than it now has.

From opposite Petro Alexandroskiya up to Tunuklu, where the Khivan oasis may be said actually to commence, and which is situated at about one hundred and sixty-five miles above Khodjeili the banks and channels of the river continue to preserve their previous aspects. Ascending, however, along this portion in which the irrigation canals commence to be diverted, the great breadth possessed by the Amú gradually contracts, while the large alluvial islands diminish in size. The cultivation in Khiva remains much the same as before, while the strip on the Russian side gets narrower and narrower until it ceases altogether at a point opposite to Tunuklu, when the desert replaces it and continues along the Amú on the right bank.

From Toyuboyin, a distance of about one hundred and ninety miles from Khodjeili, down to Tunuklu, the Amú flows in the shape of a U round a promontory projecting to the south, and either bank is lined with continuous low clay hills, covered with sand. Those on the Khivan bank are the highest, especially along the lower half of the distance, where a ridge of argillaceous schist has been cut away by the stream into a steep precipice about forty feet in height. This is terminated at Tunuklu in a rounded elevation crowned with a battery, which offered some resistance to the passage of the river by the main Russian column marching on Khiva in 1873.

At this point the Amú-darya is contracted to

a breadth of about two thousand feet, which continually diminishes up to Toyuboyin; from whence the river runs with a high velocity and with a rippling noise over the pebbles and stony *débris*, brought down by the stream and deposited along the shores.

There can be little doubt but that this portion of the Amú is referred to by Kiátib-chelebi, a writer of the 16th century, in a quotation from Hamdalla, the Arabian geographer, who wrote in the 14th century. After describing the upper courses of the Oxus he goes on to say that 'it enters a narrow valley called the Lion's Mouth, which has scarcely one hundred cubits of breadth. . . . The pass just spoken of is not far from Kourgenj (Khiva), a town of Khwarezm, and when it issues from it the Jihún loses itself in sands of two farasangs (eight miles) of breadth, where one sinks so much as not to be able to walk.' The portion of the river above Tunuklu is still known as the Lion's Mouth, though the Amú-darya has evidently enlarged its breadth considerably since it first flowed in this direction to lose itself in those sands which probably provided the material for the alluvial islands now found near Shahbázwálí.

Ascending the Amú from Tunuklu, Sultan Serai

is reached on the Khivan bank, at the bend of the U; and here the centre of the river is occupied by a pretty wooded island, covered with dwarf elms and willows of size, whose seeds must have floated down from a far distant glen in Badakhshán. The island is called Aralchí-baba-auliya, and its boskiness forms a pleasant sight, as the eye passes over it from the bare and sterile clay ridges which line the banks of the river. At Sultan Serai the native boats take in and discharge cargoes of goods, forming the trade between the Khivan Khanate and the countries on the upper courses of the Oxus. Behind the clay ridge on this side is the town of Pitnak, whose environs are watered by small cuts conveying water from the river at suitable low points through the high ground.

At the top of the reach, above the narrow gap through which the Amú rushes at Toyuboyin, the Camel's Neck comes in view. Here the river flows with a breadth of eleven hundred feet and a velocity of seven miles an hour during floods. On the right bank the limestone-rock, through which the river passes, has a vertical face of from ten to twenty feet in height; while on the Khivan bank is a higher elevation, whose perpendicular side is, as it were, roughly built up of large cubical masses, into which the rock has been shattered by some great convulsion of nature. Above the gap the Amú is seen flowing down in a channel that is bordered by ridges of low, rounded hills, all but bare of vegetation and sprinkled with sand, which extend some thirty miles upwards beyond the frontier of Bokhara.

# CHAPTER XXII.

## THE ANCIENT OXUS.

The countries of the Oxus and the early Aryans—Agriculture in Western Asia, from semi-mythic times—The Persian domination—Deserts east of Caspian since antiquity—Deposits of sand by Oxus waters—Primary flow of Oxus—Northern arms of the Araxe—The Tchardjui-Balkhán course—The Attrek course—Causes of Parthian strength—Change of Oxus towards Hyrcania—Change to the Aral outlet.

WHATEVER may have been the actual direction which was followed by the pre-historic Oxus, to an outlet in the fresh-water Asiatic Mediterranean, a glance at the map will show how admirably the great river and its affluents were disposed for the fertilisation of the plains lying to the west of the highlands of Central Asia. Solitude, aridity, and desolation have been the distinguishing features of the wildernesses east of the Caspian, according to all records of history; yet, through the mists of antiquity, glimpses are caught of more cheering and bounteous aspects of Nature. In remote ages equatorial winds may have carried moisture to Western Asia, from the broad expanse of Central African seas, instead of the dry air which now blows from burning Sahara sands. The banks of the Oxus may have, at that time, been covered with flower-enamelled pasturages, where the posterity of Japhet led the lives of shepherds and tended their flocks and herds. Increasing in numbers, the ancient Aryans would have brought cattle under the yoke, and have learned to labour the earth, with the aid of the vivifying waters which were bountifully supplied by rivers flowing down from the primeval Paradise. Still later, as agriculture developed and population grew denser, as towns were founded and the battle of life grew fiercer, there would have commenced those early migrations by which the historical civilisations of the East and West were brought into being.

Passing on to semi-mythic times, indications are not wanting that the plains of Western Turkestan continued to be a field for the agricultural labours of mankind. To the half-divine Semiramis, who erected her pillars of victory on the banks of the distant Jaxartes, historians tell us were attributed the construction of the ancient hydraulic works in this part of Asia. It was she who forced water to flow according to her will, and rendered the earth fertile and smiling, in refreshing it with moisture. With the ascendancy of the Persians we know that agriculture was the chief employment of the subject populations of Western Asia; and in the passage of the *Thalia* of Herodotus, regarding the plain in Asia through which the mighty river Aces ran, there is sufficient proof that the well-being, if not indeed the very lives, of five nations inhabiting the banks of the Oxus were dependent on the supply of water for purposes of husbandry. Polybius also hands down the information of a former epoch, and tells us that the Persians gave the usufruct of irrigated lands to the fifth generation of those who had originally placed the water upon them; and he adds, moreover, that the people spared neither labour nor expense in availing themselves of this privilege. In considering such ancient records of agricultural industry, it can scarcely be doubted that the waters of the Oxus and its affluents have been continually employed in fertilising the soil since the dimmest days of antiquity, and the causes of the sterility which characterises these regions consequently demand some cons deration.

Even the very earliest historical records speak of the frightful deserts and dismal solitudes which environed Hyrcania and stretched into Asia from the shores of the Caspian. The heaping up of sand hills by the wind, the want of water and the obscurity of the scorching air are phenomena described by the ancient historians, in terms which find their counterparts in the journals of all modern travellers, and seem, more especially in the glowing pages of Quintus Curtius, to read like a record of experiences made during some recent Russian raid upon the Türkomans. And the question is, how these seemingly ever-sterile, brigand-haunted wildernesses could formerly have been the scene of a flourishing agriculture, which supported swarming and peaceful populations? The answer appears to be that the desert aspects of these regions are the consequences of that agriculture, and afford a reflex proof that here men tilled and watered the soil in times regarding which we possess but the merest echoes of tradition.

Beneath the surface of the pre-historic sea which covered the lower parts of the plains to the east of the Caspian, deep masses of sand were doubtless deposited in hollows, whose subsequent drying up, from evaporation, has left the marked saline accumulations which are often met with. But these are nearer to the Caspian shore, and, wanting as such places are in vegetable and animal life, they possess aspects which have very appropriately been described as those 'of death.' About the level named, the Aralo-Caspian steppes have the appearance rather of a wilderness than of a desert. They are wastes, where herds of wild asses wander and find sufficient herbage for subsistence. Deprived as these deserts now are of moisture, there are ample indications that at former epochs water was abundantly distributed over their clay and gravel surfaces. Abbott in his journey from Merv to Khiva, travelled through the heart of these wilds, and described them as being broken into very irregular surfaces which were dotted with several kinds of bushes, besides wormwood and the camel-thorn. Nearer Khiva, the desert is ploughed into ravines and ridges, whose north and south directions give the idea of abandoned watercourses, reported by tradition to be ancient channels of the Oxus. The ridges are gravelly, but there is no want of sand, and water is met with, though at long intervals only, and is usually of brackish quality. Between Merv and the

Oxus, Taylour Thompson speaks of a more plentiful vegetation and of a coating of verdure which covers the desert in spring and gives it the appearance of a rich sward, extending on all sides to the horizon. Nor is there a want of tamarisk thickets or of scrub forest; and one kind of tree is mentioned as growing to a height of fifteen or twenty feet. On the western confines of the desert of Khwarezm, near the dry Uzboy bed of the Oxus, there is by no means a scarcity of vegetation; and in the early part of the year, numbers of Yomut Turkomans find ample pasturage and water for thousands of camels, horses, horned cattle and sheep. Later on in the summer, this herbage is dried into tinder, and Vámbéry describes how a spark, accidentally dropped upon it, will cause a conflagration whose flames sweep like wild-fire across the far-stretching steppes, and are only arrested on the banks of some lonely salt-lake. Besides the more or less sandy tracts, there are equally extensive areas of smooth, bare, sun-baked clay; and generally speaking it may be said, that there is far less sand distributed over the surface of the country than is generally supposed.

At the present time a volume of 2,000 cubic yards per second is carried by the canals derived from the stream of the Amú-darya over the cultivated surface of the Khivan Khanate, and spreads thereon the sand contained in suspension by the waters. What proportion such solid arenaceous matter may bear to the whole volume of water, cannot as yet be stated with precision, though detailed observations now being conducted by the *employés* of the Imperial Russian Geographical Society will soon afford information on this point. However, since the stream of the Ganges carries $\frac{1}{856}$th part, while the Indus carries $\frac{1}{800}$th part, it may safely be asserted that the canals of the Amú do not carry less than $\frac{1}{1000}$th part.[1] In every second of time therefore two cubic yards of sand are distributed over the oasis of Khiva, and for the whole period of (say) two hundred days during which the canals are open, the quantity of sand deposited will amount to 34,560,000 cubic yards, which is equal to a yearly deposition of sand to a depth of about eleven yards upon one square mile of ground.

Now that portion of the waterless plains east of the Caspian which could be conveniently fertilised by canals derived from the main stream of the Oxus, has an area of about 50,000 square miles, so that,

[1] A rough experiment of my own gave $\frac{1}{750}$th part.

with the same quantity of water which is used to-day, the whole of this area would be covered yearly by sand to a depth of 0·00025 yard. In other words, four thousand years only would be required to accumulate sand one yard deep over these regions. We have, therefore, very strong grounds for asserting that the great sandy wastes east of the Caspian have actually been caused by the distribution of the Oxus waters over the surface of the country—a process there is entirely independent evidence for supposing has been in action in these localities since the earliest times. And those who casually speak of 'a sea of sand,' describe perhaps with more accuracy than is suspected, the physical aspects of such deserts where are observed the accumulated deposits made by water arrested in its flow, and which are generally hidden from view beneath the dancing blue waves of ocean.

After the rupture of the Bosphorus, and the consequent draining off of the Asiatic Mediterranean, the level of the water remaining in the Caspian basin would have speedily fallen, owing to the evaporation from its surface having been in excess of the supply received from its tributary rivers. At the present time the Volga contributes nearly three-fourths of this quantity, which serves merely to preserve the water-spread, at about 84 feet below mean sea level; and if, as there is good reason for thinking, the greater portion of the Volga waters flowed in ancient times into the Palus Mœotis, the present Sea of Azof, the level of the Caspian would, at such a date, have been much lower than now. It is true that in early times the Oxus had its outlet in the Caspian basin, and its waters would consequently have fed this sea. But most of its stream must have then been used in irrigation, and it was, besides, even at a very remote date, deprived of its larger affluents, Polytimetus, Arius, and Margus, whose waters were employed in fertilising territories of which the capitals, like Merv, 'the Queen of the World,' possessed populations rivalling those of the great cities of the present day. Under such circumstances the Oxus was stated by Aristobulus to be the greatest river in Asia, those of India excepted, a description which allows of some approximation being made to its actual volume. For it must have been greater than the Tigris, and supposing it discharged more than the present volume of this latter river, the Oxus would anciently have contributed to the Caspian basin scarcely more than one fourth of the stream which now passes by the Volga. Under the conditions specified, therefore, the water supply to the Caspian would have been a small one, and it might actually have dried up, so as to form two comparatively small lakes. Such aspects of the basin answer very curiously to the descriptions which have been handed down by Strabo, Pliny, and Quintus Curtius, and are very strongly confirmed by a passage in Arrian which will be noticed on a subsequent page. Meanwhile it may be added that this last author is remarkably illustrated by a local tradition, which asserts that in ancient times the Caspian basin could be crossed dryshod between Bakú and Krasnovodsk. And as a further indication of a previous low level of the Caspian Sea, some proofs also exist in the architectural remains which are found below the present surface of the water at various points along its shores. Such a former desiccated condition would have intensified the westerly flow of the Oxus into this basin, and any branches or canals which might, at an earlier date still, have carried a portion of the river in a northerly direction, would, upon the fall in the Caspian level taking place, have thus been more or less deprived of their streams.

Herodotus has described the bogs and swamps which were situated at the extremity of the northern arms of the Araxe, a river whose identity with the Oxus several passages from Strabo and other early writers leave no room for doubting. The Massagetæ who inhabited these marshy regions were tribes of those Sacæ or Yuechi, who were precipitated on Sogdiana and Bactriana by the continuous encroachments of still farther distant barbarians in the north-east of Asia. Such movements would have wrecked the early Central Asian civilisations, whose existence the Greeks were aware of before the date of Alexander's expedition, from the reports of merchants whose caravans travelled eastwards across the Scythian steppes from the marts of the Euxine, or who carried Indian or Chinese goods from Balkh along the Oxus to the Caspian, and thence by the Cyrus to Colchis. It is a fair speculation therefore whether the nine and thirty northern arms of the Araxe might not have been former artificial canals which had been derived from the Oxus at an antique epoch, for the fertilisation of a pre-historic Aryan territory, whose civilisation and population had been swept away by the irruption of the Massagetic hordes.

The arm of the Araxe which Herodotus stated to flow with a clear course into the Caspian, would have been the Oxus of Arrian that possessed a breadth of about six furlongs with a greater proportionate depth, and a rapid stream, which according to Quintus Curtius carried down a large quantity of sand. These distinguishing features of the Amú-darya of to-day had been noticed, long before the writers just named, by Polybius, who says that the Oxus enlarged by its affluents in Bactriana, flowed impetuously and with a muddy stream into the plains to an outlet in the Hyrcanian Sea. By this arm also passed the Asiatic commerce, which probably followed this line from a very remote era till at least so late as the time of Pompey. And if we hear of no return traffic on the river in an easterly direction, and if Alexander made no use of the great stream as a line of communication during his expedition into Asia, the reason may be found in those hydraulic phenomena which present serious obstacles to the ascent of the Amú-darya, even by steamboats; and which may also have existed in a more intensified form in the Caspian arm of the ancient Oxus, on account of the semi-desiccated state of its receiving basin.

Nor does the existence of such an arm of the

Oxus any longer rest on a few vague passages of ancient writers. The report of the Turkoman tribes who wander in the deserts between the Amú and the Caspian, shows that an ancient river-bed runs westwards across this country from the neighbourhood of Tchardjui towards Igdy wells, and this dry bed is stated to be filled up by sand-drifts, though it can generally be traced by the eye. Igdy wells are in the old Uzboy course of the Oxus, just where this turns to the west after running down from the north, and Russian officers have observed traces of a second ancient river-bed which here joins the Uzboy. As this second channel reaches Igdy from the east, it is in all probability identical with the one spoken of by the Turkomans.

Nor is this Tchardjui-Igdy-Balkhán course the sole line along which the Oxus has flowed to an outlet in the Caspian Sea. At some still unknown point, a branch turns off in a southerly direction towards Kyzyl-Arvat, and is said to pass through the defile between the Kopetdagh and Kurrendagh hills. It then perhaps enters the head of the Sumbar which is an affluent of the River Attrek, whose mouth is at the south-east corner of the Caspian basin. At the present day, the Attrek is an insignificant stream, that is yearly carrying less and less water to an outlet. It is described as flowing in the centre of a ravine, having a breadth of some thirteen hundred yards and possessing elevated and terrace-like banks.

The former passage of some portion of the Oxus by the bed of the Attrek tends to clear up the doubts that were entertained by ancient geographers regarding the identity of the Ochus with the Oxus, and points to the Attrek as being actually the first-named of these two rivers. Such a flow of the Oxus waters by the indirect Ochus or Attrek course, could scarcely have been a natural one for them to take from near Tchardjui to the Caspian; and their diversion in a southerly direction would consequently have been due to an artificial canal, which conveyed water during the cultivating season for the fertilisation of Hyrcania. Throughout this part of the year, Ochus and Oxus would have been identical, while for the remainder of the year Ochus would have had a distinct course from Oxus and more to the south of this river; both of which conditions are named by Strabo, who, however, as well as other geographers was insufficiently informed of the reason of the variation.

It may be remarked that the former flow of the

Oxus, through the defile into the Attrek, is strongly suggestive of the origin of the story in Herodotus regarding the great river Aces, which artificially watered the lands of the Chorasmians, Hyrcanians, Parthians, Sarangians, and Thamanæans.

Dr. Rawlinson has remarked that 'the passage of the Oxus waters by the Attrek course would explain much that is obscure in the history of Parthia. The seat of this nation near the south-east of the Caspian was much too confined and poor a country to supply the vast resources, in men and means, which the Parthians displayed when they carried their conquering arms to the Mediterranean, and disputed the empire of the East with the Romans. But if in times when the Oxus debouched by the Attrek, and if a vast tract of what is now sandy desert was then a well-irrigated, fertile, and populous region, forming a part of the Parthian territory, the mystery would be explained.'

If, as has been stated, the main stream of the Oxus ever adopted the Attrek course as a line of semi-permanent flow, such a change would have been due to the same causes as those which have been investigated on previous pages. The Tchardjui-Igdy-Balkhán course would have been deserted in its lower part in consequence of the diversion of the Oxus waters for irrigation in the upper courses and of the resulting depositions of sand made in the channel of the river below the irrigation canals. Of such a change into the Attrek bed. we are not entirely without some information; for, in the second century B.C., Polybius describes how the tribes living between the Oxus and the Tanais were able to make incursions on horseback into Hyrcania, across the former river, which he states to be a great and navigable stream. His account plainly shows that the direct east and west course of the Oxus to an outlet in the Caspian, which should have been a barrier against such incursions, was then without water, and that the stream was lost in the sand, precisely in the way Polytimetus is described to end by Strabo, Quintus Curtius, and Arrian. At the time Polybius referred to, the main stream of the Oxus must therefore have been passing by the southerly Attrek course; for in such a way only could a passage for the barbarian marauders, coming from the north on horseback, have been opened into Hyrcania; as far, at least, as the right bank of this latter river.

Notices of the phenomena of change which preceded the radical transfer of the flow of the Oxus from the Caspian to the Aral basin, are wanting in the classical historians. Still it is to be noted that the information given by Herodotus, regarding the swampy region formed by the northern arms of the Araxe, is somewhat supplemented, five centuries later, by Strabo, who assigns a flow by these northern arms into the Boreal ocean, which was situated beyond the swamps inhabited by the Massagetæ. The actual mouths of these northern arms we know, from Pomponius Mela, to have been in the Scythic Gulf of the Caspian, *i.e.* in the waterspread formed by Lake Aral and its overflow to the south-west. This author mentions no Hyrcanian outlet for the Oxus, and it must therefore be inferred that the river becoming more and more embarrassed in its westerly flow, had, in the first century of the Christian era, adopted a northern direction for its main stream. Thus may have been formed the inundation which a hundred years subsequently was spoken of by Ptolemy as the *Palus Oxiana,* whose position seems to have been in the neighbourhood of the mouths of the Polytimetus or Zarafshán. About this locality the country is now occupied by deep sandy deserts which extend along a northerly band towards Lake

Aral. They were crossed by the Head Quarter Russian column of the expedition against Khiva in 1873, and their aspects prove that they were formerly covered with water, whose arrested flow deposited immense quantities of fine sand over their surfaces. Two centuries later than Ptolemy, the mention made of the 'long and broad Oxian marsh' by Ammian Marcelline, seems to indicate that the *Palus Oxiana* had been drained by the fall which had taken place in the level of Aral, in which basin the Oxus had fairly established its outlet. Whether this northerly change in the flow of the river was not also assisted by the direct interference of man is a question which will be entered upon in the succeeding chapter.

# CHAPTER XXIII.

## KHWAREZM AND THE JYHÚN.

The ancient territory of Chorasmia—The change of the Oxus to an outlet in Lake Aral—The founding of Khwarezm—The Jyhún from Tchardjui northwards—The passage of Zemark and Hiouen Thsang along the Jyhún—The towns, canals, and cultivation in Khwarezm according to the early Arabs—Káth the first capital—Evidence of a change in the flow of the Jyhún in the tenth century—Al Djordjáníja the second capital of Khwarezm.

THE ancient territory of Chorasmia—which was situated in the plain of Western Asia watered by the mighty river Aces, and whose governor gave in his submission to Alexander when the Macedonian army was in winter quarters at Balkh, 'the Mother of Cities,'—was the forerunner of the present Khanate of Khiva. Chorasmia must have been situated somewhere on the lower part of the Tchardjui-Igdy-Balkhán course of the Oxus, since Arrian tells us that it bordered on Colchis and the Amazon nation, and that a proposition was made by the Chorasmians to Alexander for a combined expedition against the countries on the Euxine. Such a proposition would be explicable only in the event of the Caspian basin having presented at such a time the desiccated aspects that have been glanced at in the preceding chapter; for under such conditions the Chorasmians would naturally have been the people to supply the guides and necessary supplies, which they offered for the passage of the Macedonian army. The *route* to Colchis and the Euxine shore would have passed across the basin of the Caspian from Balkhán, and by the valleys of the Cyrus and the Phasis. Such a passage would have been possible had the Caspian level been some 1,200 feet lower than that of to-day, and, as has already been mentioned, ancient tradition speaks of the basin having been crossed in this manner. Ibn-ul-Athír, an Arabian writer of the end of the eleventh century, has a passage which places the ancient Khwarezm near the mountains of Balkhán; while Mokadassí, who wrote in the tenth century, not only shows that these localities were formerly watered by the Oxus, but gives in detail the circumstances which led to the ruin of the ancient Khwarezm and to the establishment of the present

Khivan territory. The tradition handed down by this writer is as follows :—

'Once, long ago, the King of the East was irritated against four hundred of his nobles, and ordered them to be taken to a country situated at a distance of one hundred parasangs (four hundred miles) from the cultivated territory. This country was the actual Káth, and there they were taken. . . . . When the King banished them to Khwarezm, he diverted a canal of the Jyhún in their direction, so that they might cultivate the land. The principal arm of the river flowed at that day to a town beyond Nesá called Balkhán.[1] . . . The Prince of this town arrived on a visit to this colony and found it to be composed of capable men. He was the guest of their Chief, and gambled with him, and it was the Khwarezmian who won. Now it had been agreed that the winner should have the right to keep the canal from the Jyhún open for a day and night, and the Prince of Balkhán kept his word. But when the stream was allowed to flow freely it ran with such force that it was unable afterwards to be stopped, and in this way the river received the direction it has to-day. In Khwarezm, therefore,

[1] Perhaps situated about Mullakari, near the mouth of Uzboy.

many canals were excavated and towns were built, but Balkhán was ruined. I myself heard the people of Nesá and Abiverd say that they sometimes went to Balkhán and found many eggs (*sic*). There are also many horses and oxen there, which have become wild.' In this tradition is a description of an ancient change of the course of the Oxus, exactly similar to that more modern one whose details have been followed on previous pages, from the relation of Anthony Jenkinson and from Abulghazí Khan's memoirs. The country round Balkhán was deprived of its fertilising waters precisely in the same manner as, centuries subsequently, the environs of Kunya Urgenj were turned into a desert; and the identity of the two catastrophes leaves little room for doubting that the antecedent causes were the same in each case.

Orientalists will, probably, some day be able to fix the date of this ancient abandonment by the Oxus of its Caspian outlet, which resulted in the establishment of the modern Khivan territory. Meanwhile it is a question whether Chorasmia still formed a part of the Parthian empire when the Arcasidæ were displaced by the Persian dynasty of the Sassanidæ, in the third century of the Christian era.

Since wild cattle seem still to have been found about Balkhán in the tenth century, it is not unlikely that, after the consummation of the change in the direction of the Oxus, a portion of the flood-stream may have continued for many years subsequently to have found its way into the western channels which anciently fertilised Hyrcania and the adjacent deserts. Ibn Khordadbeh in the ninth century describes the Caspian commerce as passing to Balkh by Gorjen and the Attrek *route*, which may possibly indicate the existence of water in the Tchardjui-Sumbar line of the Oxus at that date.

The country extending northwards from Lake Karakool and the termination of the Zarafshán would, from the deep sandy deposits, which have been described, have presented considerable difficulties to cultivation; and the circumstance may explain the more westerly direction given to the artificial canal, which was subsequently adopted by the Oxus for its flow. Kiátib-Chelebi, a writer of the seventeenth century, cites the geographer Hamdalla Kazwíní, of the fourteenth century, in a passage giving a good idea of the line followed by this irrigation canal. After describing the upper courses of the Jyhún, he says: 'Near Amol (*i.e.* Tchardjui) all the plains are watered with its stream, from which Khwarezm derives its fertility. After having been subjected to several depletions in the districts of Balkh and Termedh, the Jyhún enters a hilly country and penetrates a valley called the Lion's Mouth, which has scarcely one hundred arms' lengths of breadth, and is close to the village of Toumineh, a dependency of Herat. The pass I have just named is not far from Kourgendj (*i.e.* New Urgenj, probably), a town of Khwarezm.'

The narrow gap through which the Amú rushes at Toyouboyin has been described on a previous page, and its first sight suffices to evoke the idea that the river has been conducted through it artificially. When I made this remark *in situ* on August 25, 1874, the Kirghiz *moolla* who acted as a guide to our party related the tradition that an ancient ruler of the country had turned the river into its present course. The *moolla's* story was, doubtless, founded on the tradition recorded by Mokadassí, but this last was not, however, published until 1875 by the learned Professor M. J. de Goeje, who extracted it from a MS. procured from Constantinople.

Kiátib-Chelebi continues, that after the Jyhún issues from the Lion's Mouth 'it loses itself in sands having a breadth of two parasangs, where one sinks so much as not to be able to walk.' Now just below Tunuklu the broad part of the Amú commences where the alluvial islands occur. These are formed of a material so fine as to fully justify the verse in Rudeki,[1] in which the 'silken banks of the Oxus' are spoken of. It is possible that the sandy deposits in this locality are the remains of a portion of the old *Palus Oxiana* of Ptolemy, which would have had a tendency to drain itself to the west towards Uzboy. Ancient historical tradition tells us that the site of Bokhara was a swamp, and the lakes at the ending of the various arms of the Zarafshán, were still favourite localities for quantities of water-fowl at the beginning of the thirteenth century. There are many traces of drainage channels near Shurakhán on the right bank of the Amú, which run in an east and west direction, and there is also a large dry river-bed which follows the 42nd parallel of latitude, along the southern foot of the Shaikh-jaili range. Ibn Haukal, writing in the tenth century, describes the Oxus as crossing another river, which may, probably, have flowed in the dry bed just named. The possession of a regularly-defined

[1] This was the first poet who used the modern Persian dialect.

channel by the Amú below the 'Lion's Mouth,' as well as the present greater breadth of the pass above it, are proofs that the river possesses a larger volume of water at the present time than it did in Hamdallah's day.

From the sands, Kiátib-Chelebi says, 'the Jyhún reappears in Khwarezm, a province in which it divides into several canals . . . . all of which are navigable, and carry boats to the Lake of Khwarezm.' Here is a rough sketch of the Khivan Khanate, much as it exists at the present day, and the accounts of other Oriental authors for six or seven centuries previously show that the modern Khwarezm has always been situated chiefly on the left bank of the Amú-darya below Toyouboyin.

The earliest notice of Khiva by a western historian is generally considered to be that contained in Menander and included in the account of the envoy Zemark, who was sent in the sixth century by the Emperor Justin to Djizabul, the Grand Khan of the Turks. Zemark, apparently, stayed in Káth, the old capital of Khwarezm, on his return journey to Europe; and, after having crossed the Oich (Oxus), passed by the fortified towns of the country upon the left bank before he arrived at the 'great and wide lagoon' of Aral. The next notice of the territory seems to be that of the Chinese traveller Hiouen Thsang, who in the seventh century describes Khiva as forming but a narrow band of cultivation upon both banks of the Oxus.

In the ninth century Ibn Khordadbeh and Jakubí speak of Khwarezm as being situated upon the lower courses of the Jyhún, which entered the north-east portion of the united Aralo-Caspian Sea; and neither of these writers speak of a western outlet for the river. From subsequent Arabian historians we learn that Khwarezm was watered by artificial canals, derived chiefly from the left bank of the Jyhún, and along the courses of these were numbers of populous towns, which were generally fortified by surrounding walls and moats. They contained public buildings, such as prisons and mosques, besides markets where all articles of luxury and every kind of merchandise could be bought. The country is described as being highly cultivated and abundantly watered; vineyards, gardens, and orchards surrounded the towns; so that, according to Mokadassí, 'in Khwarezm the traveller passes continuously between buildings and gardens

—many wine-presses are found there, as well as fields, trees, fruits, and other good things.'

Upon the right bank of the river the cultivation was not so extensive, and there were fewer towns. About fifty miles above the capital Káth, which was situated near Sháhbázwalí, a canal called Gaokhwara was detached on this side of the Jyhún, at about twelve miles below the gap at Toyouboyin, and watered all the country down to the capital. Above the head of this canal there was no cultivation on the right bank in the direction of Amol, but on the opposite or western bank, one day's journey above Gaokhwara, irrigation commenced at Al-Tahiríja. From this place down to Hazarasp, a distance of three days' journey, the cultivation consisted of merely a narrow strip, but at Hazarasp it widened out to a breadth of a day's journey (say twenty-five miles) opposite to Káth. At Al-Djordjáníja, about three days' journey in a north-westerly direction from Káth, the cultivation narrowed again to some six or seven miles in breadth; and it finally ceased at Gít, about thirty miles in a south-westerly direction from Al-Djordjáníja. Gít was thus in the neighbourhood of the Uzboy course of the Oxus; and Istakhrí states it to have been near the mountain beyond which is the desert, *i.e.* near the Ust-Urt plateau.

Káth, the capital of Khwarezm, on the right bank of the Jyhún, was built after the destruction of a still older capital, which appears to have been situated at two days' journey lower down the river. In the tenth century Káth consisted of a deserted citadel, which the river threatened to carry away, and a town, which the river had partially destroyed; for which reason the population had established themselves farther to the east. It was intersected by canals, which were liable to be flooded by the river, and thus at a subsequent date consummated the catastrophe that caused the disappearance of citadel, mosque, and prison, without leaving any traces. At the date of Anthony Jenkinson's visit, in the sixteenth century, nothing but a fort seems to have occupied the site of this ancient city.

Descending the Jyhún from Al-Tahiríja, the first canal derived on the left bank was that of Hazarasp, which was about half the length of Gaokhwara, or nearly twenty miles long; the second was larger, being about thirty miles long, having its head some eight miles lower, and was called Kerderan Shah. The third canal was still larger, and carried boats to Khiva; and two miles below this was the canal of Medra, which appears to have been larger still, since it was twice as great as Gaokhwara and consequently more than forty miles long. In a high-flood, therefore, the stream of this canal would have passed some considerable way towards the foot of the Ust-Urt plateau, which is now followed by the drainage line of the Uzboy.

Two miles below the head of Medra was the beginning of the fifth canal, called Wadak; and at ten miles down its course the canal of Buh united with it, after having left the main stream of the Jyhún, some little distance below the point where Wadak was diverted. Below this canal of Buh the Jyhún appears to have continued, under the name of the canal of Kerder, to its outlet in Lake Aral. At the mouth there was a marsh covered with rushes, and there were no villages or houses, but only a few fishermen. Kerder was formed of four separate branches, which reunited and formed a single arm, described as being nearly as large as the united canals of Wadak and Buh, and having a length of nearly two days' journey. Balkhí-Istakhrí says: 'It is stated that the bed of this canal was formerly the bed of the Jyhún, and, in fact, the stream diminishes in this channel when the Jyhún diminishes.' Here, then, in the tenth century, we have good evidence of an impending change in the flow of the Oxus out of the Kerder canal, by which it reached Lake Aral in some other direction. That such a change was actually in process of taking place will be shown further on; meanwhile a passage from Masúdi, another writer of the tenth century, may be noted, which states that 'some people say the Jyhún loses itself in ponds and marshy places.' This description shows that the Oxus waters must at this date have been deserting the Aral basin and seeking a fresh outlet, and points to the probability of the main stream being lost on the line of the Uzboy after having recently adopted one of the canals flowing down to this drainage line.

On the united Wadak and Buh canals boats descended the distance of a day's journey to Al-Djordjáníja on the left bank. This city was the second capital of Khwarezm, and next in size to Káth. Al-Djordjáníja occupied a position somewhere in the neighbourhood of Kunya Urgenj. It was a slave-market and fur-mart, and the pastoral Ghozz Turks brought their sheep and cattle there for sale. Caravans went from Al-Djordjáníja to

Khozar-land (*i.e.* to the lower Yaik and Volga) and to Khorasan ; and it was, doubtless, the place referred to, under the name of Orna, and described as a great centre of commerce, by the friar Pian Carpin in the thirteenth century.

# CHAPTER XXIV.

## KHWAREZM AND THE JYHÚN—*continued.*

The dam on the canal at Al-Djordjáníja—Destruction of the town in the thirteenth century—Evidence from Arab geographers of fluctuations in the flow of the Oxus, from the tenth to the sixteenth century—The river Arzass of the Great Russian Map—Proof from Balkhí-Istakhrí that the Jyhún bifurcated in the tenth century—General proposition that from this date some portion of the river reached the Caspian by Uzboy at different epochs—The difficulty of a continuous flow down this course of the Jyhún.

THE united canals of Wadak and Buh, whose stream passed down to Al-Djordjáníja, would apparently have flowed on to the west into Uzboy, had not an artificial dam, which is described by the Arab historians as a wonderful work, been erected to divert the water which was threatening to carry away the town. It consequently follows that the discharge in this canal had been increasing since the date it was first diverted from the main stream of the Jyhún, whose mouth was in Aral, at the extremity of the Kerder canal. The dam which has been mentioned arrested the stream at an arrow's flight from the town, and threw it in an easterly direction towards Karjat Barátegín, 'near the desert.' This place, according to Balkhí-Istakhrí, was a fortress on the Ghozz frontier and these people occupied the Ust-Urt plateau; 'it was situated in the desert near the mountain and exported building stones.' In lieu, therefore, of the dam having thrown the waters of the canal in an easterly direction, it may more correctly be said to have turned the water to the north; but east may be used in opposition to west, towards which quarter the canal had a natural tendency to flow. The frontier fortress of Karjat Barátegín was, in all probability, situated at the extremity of the counter-fort of Ust Urt, which runs out from the plateau in a south-easterly direction. This forms the southern limit of the Abougir Gulf of Aral, and over it that ancient overflow took place from the Lake which formed the Scythic Gulf of the Caspian of the classic writers. All drainage passing to the south of the counter-fort would necessarily reach the Uzboy line and pass on to the Caspian; and the description of the way in which the dam turned the water at Al-Djordjánija shows that the increased flow in the canal would have naturally passed into Uzboy, where also the canals more to the south, and especially the one called Medra, had a tendency to discharge their surplus waters.

Al-Djordjáníja was taken by the Tartars in A.D. 1220, after a long siege, and its population was massacred. The capture was effected by breaking down the dam that diverted the water of the canal to the east (*i.e.* to the north) and the stream (which must at that date have probably increased in volume, from the continuous action of change), recurred to its natural western (or southern) line of flow, inundated the town, swept away all the buildings and turned the site into a swamp. Those of the inhabitants who escaped being drowned or crushed by the falling walls fell by the destroying edge of the Tartar's sword. In the year A.D. 1231 Al-Djordjáníja was replaced by the present Kunya Urgenj. This city was visited a century afterwards by Ibn-Batutah, who reached it after travelling in all haste during three days, from the banks of the Ulu-Su or 'great water;' which he crossed in coming from the direction of the Caspian across the Ust-Urt plateau. Kunya Urgenj was a great commercial emporium, during the fourteenth and fifteenth centuries, and was visited by Anthony

Jenkinson in the sixteenth century, during his commercial mission into Central Asia, as has been mentioned on a preceding page.

As regards the direction of the flow of the Jyhún at different epochs during the periods treated of by the Arabian geographers, the evidence is daily being strengthened of those continual fluctuations of the river whose causes have been sufficiently discussed on previous pages. It is not, indeed, too much to say that every hint bearing on the Jyhún which can be culled from Oriental records points to the existence and active working in ancient times of the same phenomena that have already been described as characteristics of the Amú-darya of the present day.

In the tenth century Ibn-dostcha, Balhkí-Istakhrí, Ibn-Haukal, and Masúdí all state the mouth of the Jyhún to be in Aral, though some hints have been adduced of a passage of a portion of its waters in another direction; and, as will presently be seen, Balkhí-Istakhrí has one passage distinctly showing that this portion must have been flowing down to the Caspian, though nothing was recorded of the circumstance at the time.

Mokadassí, in the beginning of the eleventh century, does not, however, speak of any Caspian outlet for the Jyhún, and the same is to be noted of Idrisí, in the twelfth century. Neither in the following century does Jácut, who fled across the Khwarezmian desert to Khorasan, when Al-Djordjáníja was attacked by the Tartars, speak of any flow of the Jyhún water in Uzboy, though such must certainly have been taking place. In this author, however, is a contradictory passage which gives an outlet to the Jyhún in the sea of Tabaristán, *i.e.* the Caspian sea. The fact of Ibn Batutah having in the fourteenth century crossed the Ulu-Su, at three days' journey to the west of Kunya Urgenj, is alone almost decisive of the fact that some of the stream of the Jyhún must then have been flowing to Lake Sarakamish on the Uzboy course. Nor should it be forgotten that Shums-ud-dín Dimaskí (thirteenth and fourteenth centuries), who places the mouth of the river in Lake Aral, has also some vague idea of a flow of the Oxus to a westerly outlet. He records, 'it is erroneously stated that, after entering the Lake of Khwarezm, the Jyhún, flowing from a gulf, runs westerly, to the country of Kerman, and then enters the Persian Gulf.' In reference to this, it should be noted there was an opinion prevalent in former times that the

Caspian waters communicated in some way with the Indian Ocean ; and the foundation of the error noticed by Shums-ud-dín may, therefore, have been the actual entry of a branch of the Oxus into the Caspian Sea.

Abulfeda, a careful compiler of the fourteenth century, says that, in reference to the Jyhún, 'opinions are much divided, but that of Ibn Haukal is most probable;' and he then describes the river to empty into Lake Aral. However, Hamdalla Kazwíní, writing in the middle of the same century, gives two distinct outlets to the Jyhún. In his description of the Caspian he says that the Jyhún (Amú-darya) is one of the chief rivers entering the sea ; and in that of Lake Aral he makes one part of the stream of the river to fall into this basin.

At the commencement of the fifteenth century, Clavijo, in the narrative of his embassy to Timúr, says of the Oxus : 'This river falls into the Sea of Bákú ;' while Abul Hassan Syed Ali al-Djordjání makes the Jyhún fall into 'the Lake of Abesgun, Lake of Kazars, or Sea of Gorgan' (*i.e.* the Caspian), 'where the Ittil' (or Volga) 'also falls.' In the same century we have also the MS. of a Persian writer, which is in Sir Henry Rawlinson's possession. The author of this MS. says, in a description of the rivers of Asia, 'in all the ancient books it is recorded that from this point the river Jyhún (or Oxus) flows on and disembogues into the sea of Khwarezm; but at the present day this sea no longer exists, the river having made for itself a new channel, which conducts its waters into the Caspian. . . . From Khwarezm the greater part of the country is desert, down to where the river falls into the Caspian.' Again, in reference to the Syr-darya, the same author states: 'the river of Khojend, in the lower part of its course, passing into the desert of Khwarezm, joins the Jyhún (or Oxus), and thus ultimately reaches the Caspian.' That such a junction is physically possible will be shown on a future page; while subsequent Russian information already quoted affords support to the passage that has just been cited.

For the sixteenth century we have the evidence regarding the flow of the Jyhún, from the relation of Anthony Jenkinson's journey, and from the memoirs of Abulghazí Khán, which has already been discussed; and from this it cannot be doubted that the Oxus had recently deserted its Caspian outlet, and had effected an entry for a second time into the Aral basin. Such a flow of the Oxus to the Caspian is corroborated by Russian information found in the legend of the Great Map, and mentioned on previous pages. This says that, 'from the Blue Sea' (*i.e.* Aral) 'the river Arzass flowed to the sea Khwalim' (*i.e.* the Caspian), 'and into Arzass fell the river Amú, coming from the east.' It seems probable that the expression 'from the Blue Sea should be more properly read 'from the *direction* of the Blue Sea,' and that the Arzass is identical with that branch of the Syr alluded to in passages of the Persian MS. which has been mentioned above.

In the seventeenth century, Kiátib-chelebi, citing Hamdalla, says that 'one arm of the Jyhún discharges into the Caspian Sea near Khalkal, a place situated at six days' journey from Khwarezm, and inhabited solely by fishermen. The author of Mesalek-ul-memalek, and that of Tecouim-ul-boldon, Ibn Haukal, and Abulfeda, say that the mouth of the Jyhún is in Aral, but one may believe that it is only the principal arm of the river of which these authors speak.'

A passage of Istakhrí has already been quoted, from which it appears that in the tenth century the flow of the Jyhún by the Kerder canal, that was thought to be the old course of the Jyhún, then showed indications of change. There is a second passage of the same author, not only specifying the new direction adopted by the river, but showing that the Uzboy course of the Oxus actually dates from the tenth century, in which this author wrote. This passage is as follows: 'Opposite to Gít, but on the northern bank, is situated, at one parasang (four miles) only from the Jyhún, the town of Madhmenia. This town, though it is situated on the right bank, belongs to the territory of Al-Djor-djáníja, which happened thus—because the river Kerder turns or has taken a new southerly direction, and runs between the territory of Gít and Madhmenia.'

Madhmenia was, therefore, on the right or northern bank of the new westerly (or southerly) course of the Jyhún, when the stream changed from the northern outlet by the Kerder canal; and the new course flowed between Gít and Madhmenia. The ruins of Madhmenia-kala are situated at about fifteen miles due south of Kunya Urgenj, and this new line of flow of the Jyhún was, therefore, in a canal about twenty miles due south of Kunya Urgenj, and ran towards Lake Sarakamish. After the taking of Khiva, the Russians, marching northwards, crossed the old dry bed of the Amú-darya, which is called Doudon, and learnt from the Turkomans that it passed by the hill Mongyr towards Sarakamish. Such information, therefore, corresponds very accurately with the line of flow stated by Istakhrí, in the tenth century, to have been adopted by the Jyhún. Once flowing in this direction the river had no choice but to discharge its waters into the Uzboy, by which they would ultimately pass into the Caspian, at the Balkhán Bay.

The Doudon bed was evidently one of the canals (most probably the one called Medra) derived from the Jyhún below Khiva, and answering to the course of the Jyhún alluded to by Kiátib-chelebi, in the following words: 'There is an arm of the Jyhún, which, after naving passed the capital of Khwarezm, enters a narrow and stony valley called Kerlawa by the Turks (*i.e.* the Ghozz, who inhabited the deserts towards Ust Urt). This arm then forms a cataract, where it falls with so terrible a noise that one can hear it at two parasangs' (eight miles') distance. According to Hamdalla, this arm discharges into the Caspian near Khalkhal, situated at six days' journey from Khwarezm, and entirely inhabited by fishermen.'

Although, therefore, many Arabian writers, in their description of the Jyhún, have omitted to make any distinct mention of a Caspian outlet for a portion of its waters, ample evidence has been adduced that, commencing with the tenth century down to nearly the middle of the sixteenth century, the river bifurcated and sent a large volume of water to a westerly outlet. The causes of such capriciousness of flow have been fully discussed, but in addition to these there is no reason for supposing that the changes may not have been hastened or delayed by the erection of artificial dams on the different branches, such as the one which is specially noticed to have existed at Al-Djordjáníja. It is also a question whether these phenomena of change may not have been complicated by the varying level of the Aral waterspread; which, according to its height, would have repelled or invited the flow of the Oxus into its basin, at different dates. It results from the examination that has been made of the Amú-darya that, though the river can be easily made to flow towards the Caspian, any permanency for such a flow is possible only in the event of the stream being diverted above the territory of Khiva. After passing through this country, the waters on their way to the Caspian basin would have to flow along the Uzboy, which, down to Igdy wells, traverses a country whose small slope assists the fullest development of the evils caused by the quantity of sand held in suspension during the floods of the Amú, and by the uses of irrigation to which a large portion of those waters are put.

## CHAPTER XXV.

### THE STEPPES OF TURKESTAN.

The Turkoman chief, Atta Murád Khán—The Kirghiz Safar—Kirghiz and Turkoman horses—Hard conditions of life both for man and horse—Climate—Cold Polar winds—Heat—Dryness—Dew—Rain—The *kibitka* of Asia—Kirghiz shepherds of Kizzelkoom—The safety of the desert at the present day—Fitness of Russia for civilising Central Asia.

THE presence of a steamboat on the lower Amú naturally attracted the curiosity of many Khivan notabilities, among whom was a near relation of the Uzbek Khán, who, however, was in no way remarkable. Another visitor to the 'Peroffsky' was the Yomut Turkoman chief, Atta Murád Khán, of the famous race of the Atabegs, or governors, of the Persian kings. Notorious for half-a-century of warfare with the rulers of Khiva, at the end of which he had been eventually worsted, the interest he presented was added to by the circumstance of his being an old friend of the commander of the Amú-darya expedition of 1874. This officer had formerly served in the trans-Caspian territory, and had known the Turkoman chief when he was in exile among the Yomut and Tekke *aouls*, on the skirts of the Kopet and Kurun-dagh. Though the rugged exterior of the old chief corresponded very well with the *beau-ideal* I had pictured to myself of the leader of a desert horde of nomadic brigands, the childish affection with which he embraced his Russian friend of former years suggested that even the fierce Turkomans were not so devoid of human feeling and affection as they are usually supposed to be. Another individual was of interest for a different reason. He was a tall, fine-looking Kirghiz, and a Russian *employé* on the Amú, who told me that his father had been (as I understood) the chief of Abbott's escort when this officer passed from Khiva across the Ust-Urt plateau to the Caspian in 1842. His name was Safar; and he mentioned with some pride that the British officer had sent his family some presents subsequent to arriving in England.

Having returned to Núkús from Toyuboyin, it became necessary to make some arrangements for a contemplated ride across the Kizzelkoom desert to Fort Peroffsky, upon the Syr-darya. Horses could only be procured at Cimbye, which place was easily reached in twelve hours by boat, *viâ* the Kuwán-jerma arm of the Amú and the Kigailie canal. On our arrival it was found that, with the exception of a few Turkoman horses, the sole beasts available were those of the Kirghiz breed. These seldom stand more than thirteen hands and a half in height, but their lack of size is amply compensated by wonderful powers of endurance, as may be judged from the fact that an animal which in England would be nothing more than a coarse pony, carried me and my saddle-bags—a weight, perhaps, of sixteen stone—more than thirty miles a day, for ten days in succession. Nor was he a bit the worse for the performance, though he had but very short rations of grain, no grass, and little water during the whole of the trip. The Turkoman horse, which is a lighter-built, taller, and more blood-looking animal, is in no way inferior in hardiness; both breeds make enormous distances at a 'butter and eggs to market' jog-trot; and the utter neglect with which they are treated upon long journeys gives a wonderful shock to European ideas regarding the necessity of grooming, feeding, and watering.

For the dissipation of these and others which pass on the Continent for British prejudices, a ride across the Kizzelkoom may be confidently recommended. If bread alone be not the food on which man can be nourished, the Englishman would put in a plea for a portion of beef, or at least animal food of some sort, but for the Russian or the Tartar sour black biscuit and weak tea will suffice. An individual with these limited ideas regarding nutrition accompanied our party across the Kizzelkoom, and, being the strictest of Mohammedans, confined himself to the above provisions, firmly refusing to share an occasional scanty ration of meat, which had been slaughtered by infidel hands. It is difficult to say what the secret of such immunity from hunger may be, more especially when long distances are being ridden every day, and night after night is spent *à la belle étoile*, in rain, and frost, and snow. One of our number suggested pleasantly that inconvenient habits, like that of eating, ought to be avoided when traversing deserts, since, from satisfying hunger to-day, a similar inconvenient want might be felt on the morrow! This capacity for fasting is, however, a very old story; since Pliny related that ten days' journey beyond the Borysthenes, the Sarmatians take food only at intervals of three days.

A hard life of this nature seems, oddly enough, to have few ill effects on the health, for our party of some twenty men had not a pain nor an ache of any kind during twelve days of double marches between the Amú and the Syr. Notwithstanding the extremes of the West Turkestan climate, it must certainly be considered a healthy one, and from June to September, in spite of Gargantuan melon eating, the sick list on the Amú among the Russian troops was literally a blank form. Towards the end of the autumn, when the mornings have 'a nipping and an eager air,' and when local rains set in, fever and bowel complaints show themselves, but these disappear with the hard dry frosts of winter, when the courses of the Amú and of the Syr as well as the surface of Lake Aral are frozen over.[1]

The winter climate of the Aralo-Caspian basin is severe, and the thermometer frequently stands at $-20°$ Fahrenheit—a degree of cold which is greatly intensified by the predominance of Polar winds, blowing over great stretches of frozen country. Lewchine states that 5° or 6° of cold with a north wind give the same sensation as 15° or 16° of cold

[1] The lower Amú-darya was frozen from January 1 to March 4, 1875.

without such a one; and the truth of this remark I had the opportunity of experiencing when crossing the Karakoom desert at the beginning of November. Sometimes these winds amount to hurricanes; and when such a *bourrane* (as it is locally called) blows, the *kibitkas* of the nomads are carried away, and their cattle are driven for miles before the wind, to perish in the snow. Of 12,800 camels which accompanied Peroffsky's expedition in 1839-40 only 200 escaped; but this, it should be noted, occurred on the Ust-Urt plateau, where a winter temperature of $-30°$ Fahrenheit is nothing remarkable. The great depression of the Aralo-Caspian deserts, the want of water, and the all but entire absence of vegetation, are circumstances which cause the great extremes of heat and cold. As the case of Egypt indicates, nothing but the reclaiming of these vast stretches of bare waste land will ever modify this evil. Though planting has not, in the country just named, as is generally supposed, increased the rate of rainfall, the development of cultivation by the waters of the sweet-water canal has perceptibly ameliorated the climate, by tending to equalise the summer and winter temperatures. When work on the Suez Canal first commenced, the thermometer on two occasions descended below freezing point—a degree of cold which has not occurred since, and will not, probably, occur again. The change has been attributed to infiltration of moisture by capillary attraction and is, doubtless, in great measure due to the vegetation which has sprung up wherever the stream of the sweet-water canal has been applied to the soil.

The sufferings of the Djizzak column during its advance on Khiva, across the Khalatta desert, in the spring of 1873, will give a good idea of the climate of the sandy deserts of Oxus countries at a season when the summer has not yet fairly set in. The shores of Annesley Bay were recalled to my recollection on the hottest day I experienced on the Amú in 1874, when the thermometer reached 110° Fahrenheit in the shade. The scorching air was blowing from the Kizzelkoom on the east, and, carrying with it an immense quantity of fine sand, invested the landscape with a gloominess almost like that of a London fog in November. It is at such a moment that the view of the fresh green Khivan oasis on the opposite bank fully answers the wildest flights of Oriental hyperbole.

In the Aral basin itself very heavy dews occur, which are not noticeable outside the basin. The circumstance is noted since rates of evaporation determined by observations at the Russian observatory at Núkús will give erroneous results when applied to determining the loss of water from the surface of the Lake.

Such limited quantity of rain as falls in these latitudes is brought almost exclusively by a south-west wind, and is frequently evaporated before reaching the ground. About Lake Aral rain is frequent during the autumn, though little drainage runs off the thirsty soil; but on the Lower Amú in 1874 two exceptional days of heavy rain occurred. On this occasion our party found shelter in a *kibitka*, which was carried bodily across a swollen stream to us, by a dozen brawny Karakalpaks, whose heads only showed above the surface of the water.

Of all temporary habitations which are welcomed by the weather-worn traveller, none can come up to a *kibitka* for snugness and for comfort. Sheltered behind its felt-covered walls from the howling wind outside, and basking in the flames of the fire, whose wreaths of smoke rise gracefully to the opening in the roof, the dreariness of the desert is forgotten, and fatigue soon passes, as, wrapped up in lambskins, one blesses, with honest Sancho, the man who first invented sleep. For the carriage of a *kibitka* a good camel is required; but a small party wishing to travel speedily across Asiatic deserts can usually procure one from the nomads, who are encamped round the successive wells, where a halt must be made. The Russians make extensive use of *kibitkas* for sheltering their troops in Turkestan and on the east shores of the Caspian; and about fifteen men can be comfortably accommodated in one, weighing about four hundred pounds, its felt coverings included. Besides those of the usual beehive shape, there is one of a smaller and lighter pattern, the ribs of whose conical roof radiate from a central point, where they are connected like those of an umbrella; and this roof is carried in a single package, while the walls form another. As regards the antiquity of the *kibitka*, there can be little doubt that this abode is referred to by Herodotus in the passage where he describes the Argippæans as dwelling under trees (frameworks?), which they cover in winter with cloths of thick white felt.

As the memory recalls the manstealing and rapine and murder that have characterised the Turkestan steppes for ages, the solitary *kibitka*, which looks like a speck on the far-stretching expanse of the desert, must forcibly excite the dullest imagination. Nor does it diminish the feeling of mingled wonder and pity with which a European asks himself how this nomad life can be bearable when, on riding up to such an abode, it is found tenanted, perhaps, by a couple of women, and a troop of merry-looking children. The good man is miles away, shepherding his goats or his camels, at some remote spot affording pasture; while his household and his belongings seem to be at the mercy of the first bloodthirsty ruffian who may happen to pass. But times are now changed; for, though the redoubtable brigand chief Sadyk still lurks in the more remote deserts of Bokhara, solitary shepherds and defenceless women, no less than merchants with well-guarded caravans, may roam unconcerned, or live in safety in the wastes of the Kizzelkoom. The days of Turkoman and Uzbek *alamáns* are over, and the Kirghiz, whose poverty drives him to such distant pastures, need no longer sleep with half-open eyes and his sword in his hand, on guard against the plunderers who formerly stole his women and children in lieu of other tangible booty.

Hospitality is the virtue of the desert, and the passing horseman will always find a welcome and a bowl of milk at the meanest *kibitka* in the deserts of Turkestan. And as he hands back the emptied vessel to the Kirghiz mother, who has one infant in her arms and a second clinging about her knees, and goes on his way re-echoing the *Amán o Khúsh*—peace and happiness—which forms her farewell salutation, he is able to judge of the benefits that have already been conferred by a civilised government on these poor people, in making the phrase an actuality, instead of the empty formula it used to be in the past.

This is one phase of the consequences of Russian advance in Asia, which, not unnaturally, but too often, escapes remark by critical people, well guarded by stalwart policemen, and unacquainted with a social state that is solely capable of being directed by the rough-and-ready justice of the strongly-armed hand. Those who often deplore the low type of the civilisation which Russia introduces into Central Asia may at least take comfort in the reflection that everything is for the best in this best of all worlds, since she alone, perhaps, among the great European Powers, is capable, from the very simplicity of her social institutions, of conferring on these Asiatic nomads the very benefits which as yet they are able to comprehend and can feel grateful for.

# CHAPTER XXVI.

## A RIDE ACROSS THE KIZZELKOOM.

Country east of Cimbye—Canal of Púr Khán—Fort at Kaleech-kala—The Kirghiz chief Nazr Khán—Baggage animals—The Belitao hills—The low ground beyond them—Ancient lake in this depressed furrow—Present aspect of country—Evidence of volcanic upheaval—Caravans and shepherds—Buzganak—A comfortless bivouac—Reflections round a camp-fire in the Kizzelkoom.

THE route from Cimbye to Kaleech-kala, situated some fifty miles to the east, in the low marshy ground immediately to the south of Lake Dowkara, soon leaves the cultivated fields, and enters on a boggy tract, clothed with tamarisk jungle, which is much cut up by canals and water-courses. These leave the Kuwán-jerma arm of the Amú below Kigailee, and in the spring and summer carry a good deal of water to flood the low ground. Their heads have, however, generally dams built across them by the nomads, in order to preserve sufficient grazing-ground for their cattle. At intervals upon this country are scattered a few homesteads surrounded by cultivation, but the greater part is waste, and studded with ridges of sand, between which are shallow reedy lagoons and swamps. The population is sparse, and comprises a few miserably poor Kirghiz, whose *kibitkas* are pitched on the sand-ridges at long intervals. In going to Kaleech-kala, some of the canals mentioned have to be crossed; and at the largest one, called Púr Khán, the Cossacks of our escort were by no means so handy in swimming their horses over as might have been expected.

The ground surrounding the deserted fort of Kaleech-kala is but very little elevated above the surface of Lake Dowkara in the flood season of the year. The crenelated *pisé* walls are in good preservation, and look imposing enough from the surrounding low ground and from the navigable channel through the Lake, whose ascent by the Russian steamers used, in past times, to be awaited in fear and trembling by the Uzbek garrison of the place. The neighbourhood is, to some small extent, cultivated by the Kirghiz, whose *aouls* are the head-quarters of the chief Nazr Khán, who has been mentioned on a previous page, and whom, with his two sons, we found to be well-bred in manner and altogether very good specimens of Kirghiz nobility; though it must be remarked that their features presented a Tajik rather than a Mongol-Turk type.

With Nazr Khán's people, who wander east of Dowkara towards the Kizzelkoom, bargains were made for the supply of the baggage animals required for the transport of our party to Fort Peroffsky. For this distance, of nearly three hundred and fifty miles, twenty-five shillings had to be paid for the hire of a camel, and twenty for that of a pony; the camels were subsequently found to be inferior, but the baggage ponies could not certainly have been surpassed in endurance and weight-carrying qualities.

From the low ground about Kaleech-kala the range of the Belitao hills presents a steep cliff of about two hundred feet high to the south, whose formation is that of a soft argillaceous sandstone; but the bold aspect here possessed by the cliff, gives place, as the hills trend eastwards, to a low ridge, that, fifty miles away, is little more than a high bank of clay. On the summit of the western end of the ridge is a tall *pisé* minaret, the mausoleum of a venerated Uzbek saint, which, according to local tradition, made a tour of Khiva, and diffused the odour of its sanctity for short periods in different localities, before settling down in its present conspicuous position.

It will be remarked that the line of the Belitao hills, which run parallel to the south coast of Lake Aral, at a distance of about thirty miles, would act as an intercepting dam for all drainage that might run down the slope of the country in a northerly direction towards the Lake. Any such water could only pass into Aral by being thrown off round the western extremity of the hills, since the country slopes down also in this direction from the dry Lake Kouktchatengis on the east. Such drainage-water would not, however, reach Lake Aral immediately, for there is, along the southern foot of the hills, a depressed furrow, whose depth is proportionate to the general height of the ridge; and the water flowing from the east and from the south would collect in this low ground and would form a lake. At the present day the deepest part of the furrow, that lies under the highest part of the Belitao—*i.e.* under the cliffs at the western extremity—is occupied by the Dowkara Lake, from which the ground slopes upwards and eastwards to the dry Lake Kouktchatengis.

Shums-ud-dín Dimaskí—a writer who has already been quoted from—tells us (on the authority of two other ancient Oriental authors) that before Sogdiana was cultivated its lands were flooded by the waters of a river, which fell into Lake Aral on the east. At the outlet the waters formed a lake, twenty parasangs (eighty miles) in length and of different breadths, down to five parasangs (twenty miles), and one part of this lake was said to be in the Jyhún or Oxus.

From the above information, and from the present aspects of the localities, it may be inferred that, at some not very well-defined historical epoch, a lake, connected with that of Kouktchatengis, existed to the south of the line of the Belitao; and that this waterspread was fed by the waters of the Amú and of the Syr, of which latter river a portion also passed into Aral. The area of drowned lands so formed may have been the outlets of those northern arms of the Araxe that Herodotus speaks of, and at a later date may have received the main, or, at any rate, that part of the stream of the Oxus which remained to flow to a northerly outlet after the irrigation canals running through Khwarezm to the west had been provided for.

From Kaleech-kala the route to Fort Peroffsky runs north-east for about twenty-five miles, and generally follows the line of the Karakool channel, which, from a point south of Lake Dowkara, flows out of Kuwan-jerma, to lose its waters in the low-lying sands that are under the Belitao hills. At the commencement of Karakool a few plots of cultivation are passed through, but these soon cease, and a plain thinly covered with small tamarisk-bushes is entered upon, where bare sand-hills are also encountered at intervals. The country gradually becomes drier, and sand is more common; while *saksaoul*-bushes replace the tamarisk, and become large enough and numerous enough to be styled a forest in one locality, about ten miles from Kaleech-kala, where a large, deep, dry river-bed occurs. The bed and banks of this are covered thickly with thorny jungle for a few miles, and then a moister, tamarisk-covered plain commences, on emerging from which the banks of the Karakool channel are struck near its termination. Here the country is drier and sandier and is covered with *sak-saoul* and thorny bushes once more.

Eastwards from the termination of the water in Karakool, a dry river-bed, of some breadth, but which appears to be filled up with deposited earth, is followed east through a scrubby bush vegetation, which covers the banks and bed of the channel. Sand then increases along the *route*, while vegetation decreases, until the path enters a ravine whose clean-cut banks have a height of some thirty feet, and whose breadth is about three or four hundred feet. From the deep sand covering the old river-bed in this place the route mounts to the bank, and emerges upon a wide, open plain stretching to the horizon on the right hand, but enclosed on the left, at two or three miles' distance, by the low ridge of the Belitao hills. The hard clay surface of this naked clay plain has scarcely a crack deep enough upon it in which a seed might find shelter and germinate, and is certainly the desiccated bed of a former shallow lagoon. Here and there is a small tussock of scrubby-looking grass, or a clump of low, scattered bushes; but the want of birds, and of life generally, stamps this desert steppe with a special air of solitude and of dreariness. Across it, at a distance of about sixty miles from Kaleech-kala, a halting-place is reached, called Buzganak, where a few small ravines or dry nullahs terminate in a shallow hollow, filled with a pool of reddish water, like those to be seen in deserted brick-fields. So thick was this liquid with fine clay that our tired horses refused to drink of it, though we ourselves used it for tea.

Before reaching the muddy pool at Buzganak the track across the plain approaches the Belitao, at a point where it is nothing more than a bank of clay of about sixty feet in breadth at the base, and thirty feet high, and of so regular a form as to look quite artificial. In the line of the hills at this spot is a gap of about fifty feet in breadth, suggesting in a remarkable manner the similar cut in the Ullah-bund, on the lower courses of the Indus, a description of which will be found in Lyell's *Principles of Geology*. It is, indeed, scarcely possible to doubt that the Belitao ridge is due to upheaval phenomena similar to those which threw up the Ullah-bund, and it seems very probable that the gap mentioned as occurring in the hills was subsequently made at a weak point in the ridge by drainage waters pent up behind it.

Between Kaleech-kala and Buzganak there are a few scattered Kirghiz *aouls*, and a mud mosque or two is seen nearer Lake Dowkara. As we passed along the route a considerable number of camels were encountered, which were carrying ironwork, piece-goods, woollen cloth, and other Russian productions to Khiva. We rode up at sunset to Buzganak, where a flock of some hundreds of goats had just finished watering and were moving off homewards, with serried ranks, to some secluded Kirghiz *aoul* in the neighbourhood. The shepherd himself, who was mounted on a stately shaggy camel, stopped to hear the news we had brought from the more civilised latitudes of Kaleech-kala and Cimbye; but he followed his goats as night set in, leaving us to a dreary bivouac, which was made still more comfortless by the howling wind blowing about the scanty brands of our camp fire.

Wrapped in his fur *shuba*, each one laid down on the hard, baked ground, to seek 'tired nature's sweet restorer,' after the fatigues of the past day. For me, sleep was impossible, amidst that group of recumbent forms, lit up by flickering flame. The thoughts of an Englishman could but picture how the noble Conolly probably couched by this desolate pool, as he travelled by this *route* to find a martyr's aureole at Bokhara; and as the hours of night passed by, my mind reviewed the impressions that were given by the unaccustomed scenes I was passing through.

The gloom of the West Turkestan steppes, which first impresses one so forcibly in the Karakoom deserts north of Aral, seems surpassed by the sadness of the Kizzelkoom, near the south-east corner of the Lake. As the vision passes from ghastly-looking ridges of sand that are sprinkled with funereal-looking bushes, over immense stretches of lifeless-looking, bare, clay plains which are lost in a low continuous line of elevations on the horizon, the feeling of novelty aroused by such surroundings is almost overcome by a sense of mysterious awe. There is, indeed, something touching in the sight of a little group of mounted men, who look like specks upon the wide solitude across which they are advancing, and who seem, by the steadiness of their movement, to dare the spirit of Immensity, hovering, as it were, menacingly around them, in thick clouds of fiercely-drifting sand. The reverberation of ringing hoofs upon the hard, glazed ground, and the metallic clattering of arms, are sounds that serve but to intensify the surrounding stillness, while the few hushed words of a trooper addressing his comrade, strike the dweller in cities as unnatural in comparison with the busy hum of men to which he is habituated. It is at such a time and place that a vague yearning arises for the rustling of leaves, the rippling of waters, the chirruping of birds, or any of those small common sounds which betoken life, and, if unheeded, are at least regarded as the necessary and inalienable possessions of man and of his existence.

But as verst after monotonous verst is ridden over and the weary limbs are stiffened with fatigue, the mental faculties have a special pleasure which is all their own. The imagination may revel in history, and vivify the environing solitude with the stirring episodes of the past, when the conquering Macedonian swept, meteor-like, before the astonished gaze of the barbarians of the Oxus and Jaxartes. The successive waves of Mongol invasion, the never-ending conflict between Iran and Turan, the Arabian conquest of Central Asia, and the embittered hate of Súnní and Shíah, are mines of story whose dramatic wealth the most vigorous or prodigal fancy runs no risk of exhausting. And, passing from the early civilisations of these localities and the oceans of bloodshed which have swept them away, to the contemplation of the desolation that has taken their place, the thought can but dwell on the greatness of the task which it has fallen to the lot of

Russia to execute. Surely, if there is anything that could satisfy a noble and a boundless ambition, it is the rehabilitation of Nature, and the effacement of those marks with which barbarism has disfigured her face, in the venerable regions included in the Western Turkestan of to-day!

## CHAPTER XXVII.

### A RIDE ACROSS THE KIZZELKOOM—*continued.*

Buzganak to Shuwakbye—Dessication of Steppe on east of the Belitao—Kirghiz at Shuwakbye—Kouktchatengis—Old mouths of Jaxartes, Syr, and Kizzel-darya—Course of Jáni-darya—Mosques and Mausolea—Old fort at Chirchiktash—Irkibye—Kirghiz of Lower Syr—Ferry at Fort Peroffsky—Conclusion.

We rode off from Buzganak in the morning through clouds of drifting sand, with the grains of which the wind half-blinded us, as we passed over sand dunes sprinkled with small thorny bushes. Beyond these the hard, bare, clay plain soon extended again, along a distance of about fifteen miles, which was intersected at right angles by ridges of sand at distant intervals. About here, perhaps, may be said to exist the extremest degree of desiccation to be found, on the country between the lower courses of the Syr and of the Amú. To the left, the ground sloped gently away to the Aral basin, and the horizon on that side was bounded by low regular ridges, while to the right the plain was lost in dark sand-clouds, wherein a stray desert antelope disappeared from time to time. Small tamarisk, and subsequently *saksaoul* bushes were passed through, until a deep river-bed was reached, when the country was covered with an almost dead forest of this latter tree, whose trunks and branches strewed the ground. For a camp-fire nothing can equal the logs of dry *saksaoul*, and the roaring flames which illumined our cheerful bivouac round the wells in the dry-river bed at Shuwakbye made a sight which was pleasant to see.

The wells spoken of are narrow shafts some eighty feet deep, which have been driven through the indurated clay soil, and which contain a limited quantity of good drinking-water. It is, however, at least fifty years since a stream could have flowed in the dry river-bed, and the desiccation which has subsequently occurred, has destroyed the vegetation which must have existed in this country since very early times. Ibn-dosteh, of the tenth century, has a passage which speaks of the mass of trees, with interlaced branches, that cover the eastern shores of the Lake of Khwarezm; and the appearance of the neighbourhood of Shuwakbye fully answers to such a description. More to the north-east, though the country is still desert, the underwood, scant as it is, commences to have more life, and its succulency measures pretty accurately the lateness of the epoch to which the waters of the Syr flowed down towards Lake Kouktchatengis. There is no doubt that vegetation where water commences would be more plentiful were it not for the goats and camels, which eat up the young shoots of the plants as fast as they are put forth. The halting-places of travellers are, of course, always at the best and most plentiful springs, and no green thing has a chance of existence, certain as it is of being immediately cut short by the powerful jaws of the camels conveying baggage and merchandise. In this way the desiccation of the country is being continuously increased.

At Shuwakbye is a Kirghiz *aoul*, the hard life of whose inhabitants is legibly graven in the furrowed and weather-worn features, that really make it difficult to distinguish the sex of an individual. During the Khivan campaign the column of troops from Cazalinsk marched by this place in successive detachments, which seem to have impressed the nomads with a sense of the benefits to be derived from an intercourse with Russians. One old man insisted on selling us what he called forage for our horses, though the animals (all accustomed as they were to hard fare) did not show much approval of the provender. There was an old woman, too, who very successfully aroused our sympathies (represented by twenty copek pieces) by burying her two children (represented by two small bits of stick) in graves made with her forefinger in the sand round the wells, where we were watching our cattle being watered.

The country crossed between Kouktchatengis and Aral is covered with tumbled sand-dunes in a crescent shape, and placed in lines running generally NNW. and SSW. upon the hard, naked clay, plain. This tract extends about twenty miles east from Shuwakbye to the course of a deep dry river, that runs towards Aral; and, after crossing this, extensive tracts of sand occur, until the old Jáni-darya bed, which conveyed a portion of the Syr waters into Kouktchatengis is reached.

From the termination of the Karakool channel up to this point, a distance of about two hundred miles from Fort Peroffsky, the country is almost waterless—the wells at Shuwakbye, in fact, being the sole good halting-place along a distance of ninety miles, though liquid of a muddy and brackish quality is found at Buzganak and at Saraboulak, some ten miles east of Shuwakbye. The vegetation in the low-lying country, behind the Belitao, consists of a few stunted prickly bushes and of tamarisk, while the larger *saksaoul* nearer Kouktchatengis has either died off or is dying. Animal life, besides a few desert antelopes, is represented by some magpies and crows, or, as winter sets in, by a solitary wild goose which has strayed from a passing flight. All desolate as these wildernesses are, there are some wandering Kirghiz who frequent secluded spots where a little water is to be had; and single travellers, mounted on magnificent shaggy trotting camels of the Bactrian breed, were sometimes encountered. Though entirely desert patches are not, perhaps, wanting, the country has been, without doubt, populated and cultivated at no very remote date, and it only requires water to be led across it once more, in order to be rendered habitable.

About Kouktchatengis would, probably, have been the mouth of that ancient Jaxartes which separated Sogdiana from the nomads, and whose outlet in the united Aral and Caspian basins was, according to Strabo (on the authority of Patrocles), eighty parasangs from that of the Oxus. 'However,' says Strabo, 'the Persian parasang has different values; some make it sixty stadia, some thirty, others forty.' If, however, it is made forty stadia, or a mean of the three valuations, the distance separating the mouths of Oxus and Jaxartes would have been three hundred and fifty miles, which is a near approximation to the length of the Aralo-Caspian shore between Kouktchatengis and Igdy, where the escaped waters of the Aral basin would have reached the old Oxus course. Near Kouktchatengis also would have been the outlet of that Jaxartes of Ptolemy which received the Demus and Biscatis, as well as the mouths of that Araxetes and Demas of Ammian Marcelline, which ran along the foot of the Sogdian Mountains into the open country, and carried boats down to the long and broad marsh called the Oxian. Here, too, was the mouth of the Syhún, stated by Idrísí, in the twelfth century, to be ten miles distant from the mouth of the Jyhún, as well as one of the two mouths of the Syr-darya which are mentioned by Abulfeda in the fourteenth century. After that date, and in the succeeding century, there can be little doubt but that from Kouktchatengis flowed the Syr, by the Kizzel-darya bed, towards the Uzboy channel of the Oxus, which also received the Kunya-darya-lik stream of this river, as is related in the following passage of the Persian MS. in Sir Henry Rawlinson's possession. 'The river of Khodjend, in the lower part of its course, passing into the desert of Khwarezm, joins the Jyhún or Oxus, and thus ultimately reaches the Caspian.' The slope of the ground from the neighbourhood of Kouktchatengis, to the Kunya-darya-lik near Kunya Urgenj, would admit of the junction of the two rivers, in the way described; and a future exploration of the country, along a line more to the south of the one now usually travelled over, will probably, lead to the discovery of the old course of the Syr which made a junction with the Amú. After quitting Kouktchatengis or its neighbourhood as an outlet, the Syr once more, at least, found an outlet by the Jáni-darya to the south-west, a direction which it again deserted about fifty years ago. It is doubtful whether the most careful study will ever succeed in disentangling the network of dry river-beds, that here cross the country in every direction—complicated as the problem is by numerous sand-dunes which were formed on the shores of an Aral whose water-spread has at various times attained all heights up to sixty feet above its present level.

As the old course of the Jáni-darya is ascended from Kouktchatengis the wilderness grows less and less sterile, while bush vegetation increases; and though little water is actually met with for some distance, pools occur at long intervals in the dry bed, while low-lying boggy tracts are sometimes seen, whereon water must have stood at no distant date, and has left a white saline efflorescence. The banks are in general cleanly cut and pronounced; and in the dry channel a growth of rushes begins, which gets denser and greener as the Jáni-darya is ascended. Along the one hundred miles nearest to the Syr the pasturages upon the country are excellent; and extensive swampy areas, subjected to flood in very recent years, are covered with tall thick growths of *arundo*, where pheasants swarm and tigers are common.

Upon the banks of this old river-course are many ancient Kirghiz and Karakalpak cemeteries and mosques, and mausolea belonging to some chieftain's family frequently occur. These, in some localities, take the peculiar champagne-bottle shape which the use of *pisé* as a building material renders desirable for the permanency of a columnar structure. Besides such remains, there is at Chirchiktash, a place some forty-five miles north-east of Kouktchatengis, an old fortress of considerable size, that is situated in the midst of a thick growth of jungle upon the left bank of the Jáni-darya. The place is a square in plan, with sides of half-a-mile long, and is enclosed by a massive rampart in *pisé* of about twenty feet in height, and of, perhaps, double that dimension in breadth at the base. From the broad summit of the rampart a good view is obtained of the jungle-covered interior, where a square citadel of equally massive construction is placed upon an eminence, which may be either artificial or natural. The fortress must have presented great, if not indeed, insuperable, difficulties to capture by an attacking force unprovided with artillery. Its age is not, probably, greater than that of the Mongol-Turk domination in these countries, and for more ancient relics the desert country immediately along the south of the Kizzelkoom will have to be examined.

A few miles before reaching Chirchiktash is Irkibye, where the routes leading to Fort Peroffsky and Cazalinsk separate. Here pass the caravans coming from Bokhara and carrying goods either for Orenburg or for Tomsk in Siberia; and in 1873 the Russians established a military post at this place to support the detachments of troops marching on Khiva from the Lower Syr. At Irkibye are a few mausolea near the small rectangular field-work, which, with its store-houses, is now abandoned, but which will, no doubt, be re-occupied, should the scheme of once more turning water from the Syr into the Jáni-darya course be executed.

In the neighbourhood of Kouktchatengis a few goats, herded by a shepherd lad, can alone find sufficient subsistence among the scattered bushes; but as the country is travelled over to the north-east the bed of the Jáni-darya becomes moister, and patches of herbage are more frequent and larger, and afford pasture for cattle which increase in numbers as the Syr is approached. There are few more animated or more interesting sights than that which may be seen at the end of a long day's journey in the bed of the Jáni-darya, where are pitched the *aouls* of a wealthier tribe of Kirghiz, who possess large flocks and herds. Here, at sunset, thousands of camels, cows, and sheep come trooping in from the neighbouring pasturages, and the whole population, great and small, turns out to assist in watering the animals, under the watchful eyes of the greybeards.

As the travellers ride up to such a place some women immediately hurry off to carry down *kibitkas* for their accommodation; and, these tents being soon erected, rich silk coverlets, embroidered cushions, and soft carpets are spread in the interiors. Nothing can exceed the warmth and heartiness of the welcome that is given by these simple people to the Russian officers and soldiers whose service is passed in these countries; and if the gay clothes and furs which are worn both by men and women imply anything, the Kirghiz of the Syr-darya are not only rich, but are by no means afraid of showing their wealth. To the south-west, though the nomads are poor and too often clothed in tatters and dirt, they attract sympathy from their frank and confident demeanour, and the practice of such hospitality as it is in their power to afford a passing traveller. There is no greater mistake than to suppose that Russia will suffer any inconvenience from Mohammedan fanaticism in Turkestan as far as the Kirghiz are concerned. The religion of these people is of a very different type from that of the

Indian Mongols, and may, not improbably, be replaced by the Greek orthodox faith at no very distant date. In matters of social intercourse there is scarcely any bar whatever between the Kirghiz and the Russian; and as I crossed the Syr in the iron ferry-boat at Peroffsky, in company with a *Yesáwul*[1] and half-a-dozen Cossacks, besides a Kirghiz married couple with their family of young children, the bearing of the party was sufficient evidence that, if the civilising of these nomad tribes is likely to be a slow process, it will at least have the merit of being a painless one, smoothed, as many difficulties will be, by the ethnic ties and sympathies which connect the two races.

---

I write these closing lines on Leman's banks, where poets and patriots have indulged their flights of fancy, where the minds of philosophers and historians have been refreshed by the loveliness of nature. Darlings and victims of fortune have roamed over these swelling hills and timbered slopes, and, gazing on the mysterious snow-clad peaks, have felt at one with the Creator. The mountain glaciers sparkling in the sun, or the soothing stillness of the deep blue lake, touch strings of

[1] Lieutenant.

harmony within each throbbing breast; while grief, and doubt, with the bitterness these bring, are chased like evil dreams before the morning's dawn. Here man rejoices in feeling and in living, for the sense of time is charmed away by the bright lines and tones of earth's fair face. Here has man, if anywhere, but small need of the sympathy of his fellow; for what is human praise or blame in presence of the rapture of such soul-stirring scenes!

But, looking on these, ill would it become me were I to forget those traits of kindness which so often gladdened me a year ago, in the dismal wastes which surround Lake Aral. Amid the depressing gloom of barren, lifeless solitudes, hard indeed would it have been, had man too withheld his sympathy from one who came as a stranger in the land. When cutting polar blasts swept over frost-bound steppes, cheering were the flames of bivouac fires, while, couched in the sand, I was chilled by rain or snow; but more cheering still was the friendly warmth of Russian hearts, whose remembrance will ever form a glowing spot among my memories of Western Turkestan.

*Spottiswoode & Co., Printers, New-street Square, London.*

London, Smith, Elder & C°. 15, Waterloo Place.

Printed in Dunstable, United Kingdom

91343209R00217